Biocontrol Agents (BCAs)

Performance, Importance and Role in Pest Management

nova
science publishers
www.novapublishers.com

Microbiology Research Advances

Postbiotics in Medical Bacteriology
Mahdi Asghari Ozma,
Hossein Samadi Kafil and
Amin Abbasi (Authors)
ISBN: 979-8-89530-559-1
eBook ISBN: 979-8-89530-652-9

Emerging Strategies in Antibacterial Drug Resistance Management: Mechanisms, Challenges, and Novel Interventions
Ajmer Singh Grewal, PhD,
Ashwani K. Dhingra, PhD,
Kunal Nepali, PhD,
Geeta Deswal, PhD and
Arun Lal Srivastav, PhD (Editors)
ISBN: 979-8-89530-060-2
eBook ISBN: 979-8-89530-140-1

Yeasts: Genomics, Biotechnology and Health Benefits
Somboon Tanasupawat and
Engkarat Kingkaew, PhD (Editors)
eBook ISBN: 979-8-89530-025-1

New Research on Mycorrhizal Fungus
Qiang-Sheng Wu, Ying-Ning Zou,
Yue-Jun He and Nong Zhou
(Editors)
ISBN: 979-8-88697-637-3
eBook ISBN: 979-8-88697-662-5

The Rhizosphere: Structure, Ecology and Significance
Somboon Tanasupawat, PhD
(Editor)
ISBN: 979-8-88697-674-8
eBook ISBN: 979-8-88697-810-0

Trichoderma: Taxonomy, Biodiversity and Applications
Michael S. Mouton (Editor)
ISBN: 979-8-88697-946-6
eBook ISBN: 979-8-88697-984-8

Microbial Technologies and Their Applications
Santosh Kumar Karn, PhD,
Anne Bhambri (Editors)
ISBN: 979-8-88697-913-8
eBook ISBN: 979-8-88697-937-4

More information about this series can be found at
https://novapublishers.com/product-category/series/microbiology-research-advances/

Daiana García
Editor

Biocontrol Agents (BCAs)

Performance, Importance and Role in Pest Management

nova
science publishers
www.novapublishers.com

DOI: https://doi.org/10.52305/XDYD5664

Library of Congress Cataloging-in-Publication Data

ISBN: 979-8-90134-102-5 (Hardcover)
ISBN: 979-8-90134-159-9 (eBook)

Published by Nova Science Publishers, Inc. † New York

I would like to dedicate this work to:
To all those who dedicate their lives to the advancement of science; those who observe, investigate, persevere, and begin again. In every dataset, every experiment, and every idea set in motion lies an act of resolve that continually propels society forward. Their work, often carried out without recognition, forms the foundation upon which future discoveries are built.
To the scientists whose efforts have illuminated the paths we travel today;
To those who have confronted unexpected results and continued their pursuit of knowledge;
To all who understand that scientific progress thrives through collaboration and the open exchange of ideas.
Furthermore, to the authors and contributors who made this volume possible. Every contribution—whether extensive or discreet—was essential to the development of this work. Your dedication, expertise, and commitment have enriched these pages and strengthened the collective endeavor they represent.
May this book stand as a testament to the enduring value of science and to the many hands whose efforts advance it, and may it serve as an encouragement for all scientists to continue pursuing knowledge with rigor, curiosity, and unwavering conviction.

Contents

Preface

It is estimated that the growth of the global population will require a 60% increase in current crop production to meet projected food demand over the next 30 years. Achieving such an expansion poses a major challenge, particularly because crop yields are continually threatened by infections caused by bacteria, fungi, viruses, and nematodes. For decades, synthetic chemical compounds have been the primary tools used to control these pests. Yet these substances often carry significant drawbacks: they can be toxic to humans and animals, disrupt soil and plant microbiomes, and contribute to long-term environmental degradation.

In contrast, the use of biological control agents (BCAs) has emerged as a promising and sustainable alternative for managing economically important plant diseases. BCAs—microorganisms or their bioactive metabolites—offer the potential to reduce the indiscriminate use of chemical pesticides, mitigating their harmful effects and, in some agricultural systems, even replacing them entirely. Their use is increasingly aligned with global efforts to promote environmentally responsible agriculture, safeguard biodiversity, and ensure long-term food security.

This book presents eight chapters dedicated to exploring the use of various microorganisms and their enzymes as BCAs against major pests affecting a range of economically significant crops. Several chapters also provide detailed analyses of the biochemical and physiological responses triggered in plants following the application of biocontrol agents, including:

i. the production, regulation, and mode of action of key phytohormones;
ii. changes in electrical conductivity as indicators of plant stress;
iii. patterns of mycotoxin production; and
iv. shifts in the structure and function of the plant-associated microbiome.

Taken together, the evidence discussed in this volume reinforces our belief that the application of BCAs represents a promising, effective, and environmentally sound strategy for managing pathogens of economically significant crops. Their implementation can substantially reduce reliance on polluting pesticides while supporting production systems capable of meeting the food demands of the future.

We hope that you find this work both informative and inspiring, and that reading it offers as much insight and satisfaction as we experienced in preparing it.

Chapter 1

The Effect of *Bacillus velezensis* EM-A8 Biocontrol Agent Against *Exserohilum turcicum* on Maize Physiology

María Fiamma Grossi Vanacore[1]
Melina Sartori[2]
Francisco Giordanino[1]
Germán Barros[2]
Aluminé Fessia[2]
and Daiana García[2,*]

[1]Laboratorio de Ecología Microbiana, Departamento de Microbiología e Inmunología, Facultad de Ciencias Exactas, Físico-Químicas y Naturales, Universidad Nacional de Río Cuarto, Río Cuarto, Córdoba, Argentina
[2]Research Group of Microbial Ecology, Institute of Earth Sciences, Biodiversity and Environment (ICBIA), CONICET, and National University of Río Cuarto, Río Cuarto, Córdoba, Argentina

Abstract

Northern corn leaf blight caused by *Exserohilum turcicum* is an important disease that affects maize worldwide. An efficient method to control this disease is synthetic fungicide application. However, this approach entails several disadvantages, such as environmental and health costs and pathogen resistance. In this context, biological control agents (BCA) represent an alternative management method based on epidemiological and ecological concepts that can be added to an integrated pest

---

* Corresponding Author's Email: dgarcia@exa.unrc.edu.ar

In: Biocontrol Agents (BCAs)
Editor: Daiana García
ISBN: 979-8-90134-102-5

management program. A BCA, *Bacillus velezensis* EM-A8, native to maize phyllosphere, was selected for its antagonistic ability against *E. turcicum*, both in vitro and in vivo, and its survival ability in the phyllosphere environment. Here, we analyse the BCA effect over maize physiological parameters to determine whether the antagonism is exerted only by direct mechanisms or if there are also indirect mechanisms involved, such as induced systemic resistance. The treatments involved two different formulations of the BCA (T1 and T2) and the combined applications of the pathogen (T3 and T4), as well as the pathogen alone (T5). Membrane stability was higher on T1 and T2 than C, suggesting that the formulations did not stress the plants. Hydrogen peroxide concentration increased on day 28 in T1 and T5. Salicylic acid and jasmonic acid were significantly higher in C, while phenolic compounds were superior on day 14 in T2 and T4. Treatments did not influence abscisic acid or indoleacetic acid concentrations. Principal component analysis revealed that membrane stability correlated inversely with salicylic and jasmonic acid. These results indicate that the biocontrol operated against *E. turcicum* involves not only direct mechanisms of antagonism but also the activation of plant defences and changes in physiological parameters involved in disease development. A pathogenicity assay was also carried out with isolates collected from maize leaves with northern corn leaf blight symptoms from south Córdoba province fields. The isolates were identified through ITS1 and ITS4 sequencing as *Exserohilum turcicum* with a media of 99.88% of identification.

Keywords: biological control, maize, northern corn leaf blight, *Bacillus velezensis*

1. Introduction

Maize is one of the most cultivated cereals in the world. In 2023, in a context of serious droughts, Argentina produced 36.1 MTn of corn. However, during the campaign 2024/2025, production reached 49 MTn (BCR, 2025). Plant pathogens endanger agricultural production since they affect the yield and storage of grains. Fungal pathogens can reach up to 20% of worldwide yield losses (McLaughlin et al., 2013). In the maize crop, Savary et al., (2019) estimated that 22.5% of the global yield losses are associated with pathogens and pests.

Northern corn leaf blight (NCLB), caused by the fungus *Exserohilum turcicum* Pass. Teleomorph *Setosphaeria turcica* (Navarro et al., 2021a) is a disease reported worldwide in maize-producing areas (Navarro et al., 2021b), causing an estimated yield loss of 2,5% (Savary et al., 2019, Welz and Geiger, 2000). However, if conditions are convenient due to changes in agronomical practices, the disease is aggravated (Formento, 2010). For example, in Argentina yield losses caused by NCLB can reach up to 40-50% (De Rossi et al., 2022), and 20-68% incidence and up to 5.5% severity was measured in a set of hybrids sown in the south of Córdoba province in the 2021-2022 growing season/cycle (Zuza et al., 2022).

Symptoms of NCLB are green-grey, elliptical, cigar-shaped lesions from 2 to 15 cm in length. When temperature and humidity conditions are optimal, the disease progresses into coalescent areas of dead tissue, reaching complete destruction of the foliage, with dark zones associated with conidial production (Kotze et al., 2019). The pathogen overwinters in crop residues (Montemarani et al., 2018) and secondary hosts such as Sorghum halepense and volunteer maize (Formento, 2010). Inoculum is spread by wind and rain (Galiano-Carneiro & Miedaner, 2017), and infection can start through both stomata and epidermis penetration (Navarro et al., 2021a). After a biotrophic phase of penetration and colonization that causes limited damage to host tissues, the fungus switches to a necrotrophic lifestyle characterized by plant cell plasmolysis, xylem plugging, and conidiophore formation (Kotze et al., 2019; Ohm et al., 2012). *E. turcicum* causes yield losses by reducing photosynthetically active area and duration, and indirectly by causing stem breakage (Kotze et al., 2019; Wang et al., 2010).

The management of this disease is based on genetic resistance, chemical control, and crop rotation. However, each strategy has its disadvantages. Genetic resistance is based on quantitative and qualitative resistance, although qualitative resistance (Ht genes) is usually unstable and race-specific, and the development of resistant hybrids can last several years. Nevertheless, qualitative resistance based on Ht genes is not being extensively used in Argentina and Brazil. Besides, fungal pathogens may acquire resistance towards resistant lines.

Another option for NCLB management is chemical control. Though fungicides are an effective tool to control leaf diseases, NCLB has a more rapid life cycle and sporulates early in the development of the disease. In fact, a single application may not be sufficient in susceptible hybrids growing in high-pressure environments. In this sense, there is a growing concern over the environmental cost of using chemical compounds in agriculture, which is leading to more strict regulations on its use. Several studies prove the toxicity of the fungicides over a wide range of living organisms, such as bees (Lu et al., 2023), earthworms (Zhang et al., 2024), soil microbiome (Sliti et al., 2024), fish (Yu et al., 2022), shrimps (Wang, 2024), and rats (Kwon et al., 2021).

Respect to crop rotations and secondary hosts removal, can decrease fungal inoculum, but this is hard to implement in the main corn-producing area in Argentina, where maize normally represents over 50% of the rotations (De Rossi et al., 2022).

An alternative to traditional methods is the use of biological control agents (BCA) to prevent diseases. The application of BCA is based on epidemiological and ecological concepts. The key components that determine any disease development include the pathogen, the host, and the environment. In this sense, foliar pathogens interact with other microorganisms that inhabit the above-ground surface of plants, called phyllosphere (Chaudhry et al., 2021; Legein et al., 2020). These interactions between foliar pathogens and phyllosphere microbiota can contribute to plant health through competition, antagonism, mycoparasitism, or by inducing a state of resistance in the plant (Legein et al., 2020).

In previous studies in our investigation group, Sartori et al. (2015, 2017a) screened the maize phyllosphere and selected a potential NCLB biocontrol agent, *Bacillus velezensis* EM-A8 (Genbank KY694464.1), based on its antagonistic effects *in vitro*, tolerance to environmental stresses, and effectiveness in NCLB control both in greenhouse and field trials. For this, our work aimed to determine the physiological mechanisms involved in maize defense that are influenced by *B. velezensis* EM-A8 (our BCA), analyzing the following indicators triggered during disease development in maize leaves: i) membrane stability, ii) hydrogen peroxide concentration, iii) phytohormone concentrations (salicylic acid, jasmonic acid, abscisic acid and indoleacetic acid and iv) phenolic compound concentration.

2. Physiological Parameters Evaluation Triggered by the BCA and the Pathogen in Greenhouse Trials

The assays were under controlled greenhouse conditions and repeated twice. Experimental units consisted of three pots with four seeds of the PW7123 hybrid each, repeated in three blocks, each time.

Treatments were combinations of the BCA grown in two formulations and the pathogen, as follows: C: control, T1: *B. velezensis* EM-A8 grown in formulate 1, T2: *B. velezensis* EM-A8 grown in formulate 2, T3: *B. velezensis*

EM-A8 grown formulate 1 + *E. turcicum*, T4: *B. velezensis* EM-A8 grown in formulate 2 + *E. turcicum*, T5: *E. turcicum*.

Formulate 1 (F1) consisted of *B. velezensis* EM-A8 inoculum grown in nutrient broth, while formulate 2 (F2) was also enriched with 5 g L^{-1} molasses and 10 g L-1 yeast extract. Both formulates were added with glycerol to modify aw (0.97aw) and kept for 24 h at 140 rpm and 25°C until 10^8 CFU ml^{-1} (Sartori et al., 2017b). The fungal inoculum was prepared by collecting spores from a growth plate and diluting them in sterile water with Triton until 10^4 CFU ml^{-1} from an isolate with proven pathogenicity towards the hybrid PW7123.

Plants were inoculated by foliar spray with the corresponding formulations once the fourth leaf was completely developed. After a week, *E. turcicum* suspension was inoculated, and plants were kept for 24 h with a polyethylene cover to reach maximum relative humidity and favor infection.

The sampling times started after BCA inoculation and continued every week for one month. Hydrogen peroxide concentration and membrane integrity were measured on fresh leaves. The remaining leaf tissue was frozen at -80°C, lyophilized, and stored in the dark until phytohormones and total phenolic compounds concentration were determined.

2.1. BCA Persistence on Maize Leaves

To determine bacterial survival and persistence on leaves, colony-forming units (CFU g^{-1}) were performed a week after inoculation with F1 and F2. For this, 2.5 g of inoculated leaves were mixed with 22.5 ml of peptone water (1%) and stirred for 15 min. Then, serial dilutions were performed up to 10^{-8}. An aliquot of 0.1 mL of each dilution was spread on the surface of the plated nutrient agar, and plates were incubated in darkness at 25°C for 24 h. For both formulated, the count obtained was in the order of 5 logarithms.

2.2. Membrane Stability

Membrane stability was assessed through electrolyte leakage. For this, 0,15 g of fresh leaf tissue was cut into 1 cm^2 squares and incubated for 2 h at 30°C in 15 ml distilled water. The electrical conductivity (EC) of the solution was measured to determine the initial EC. Then, the solutions were incubated for

1 h at 100°C to determine total EC. Electrolyte leakage was calculated as the percentage of the initial EC regarding the total EC.

The effect of single factors as well as their two-way interactions in membrane stability (%MS) was determined by ANOVA (F: 2,76, gl: 22, $p=0.004$). Furthermore, the %MS was not statistically significantly affected by the treatments (F: 0.83, gl: 5, $p=0.5311$) nor by the interaction between treatments and sampling times (F: 0.92, gl: 12 $p=0.5277$).

The analysis of variance within each sampling time revealed statistical differences between treatments on days 7 and 21. On the seventh day, T2 presented a significantly higher membrane stability (25.11%) than C (17.11%) and T1 (15.33%). On the other hand, 21 days after inoculation, T1 (41.99%) showed a significantly higher membrane stability than all the other treatments, which ranged from 22.72 to 29.9% (Figure 1).

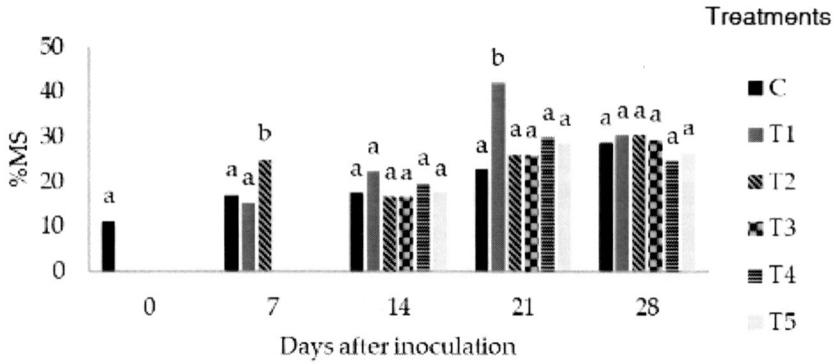

Figure 1. Membrane stability leaf (%MS) means for each treatment and time. Treatment means with the same letter are statistically similar inside each time ($p>0.05$). C: control, T1: *B. velezensis* EM-A8 in F1, T2: *B. velezensis* EM-A8 in F2, T3: *B. velezensis* EM-A8 in F1 + *E. turcicum*, T4: *B. velezensis* EM-A8 in F2 + *E. turcicum*, T5: *E. turcicum*.

2.3. Hydrogen Peroxide Leaf Concentration

For hydrogen peroxide (H_2O_2) determination in leaf, 0.2 g of fresh material was homogenized with 2 ml trichloroacetic acid (0.1% p/v) and centrifuged for 15 min at 12000 rpm. The supernatant was mixed with 250 μL phosphate buffer 10 mM (pH 7.5) and 500 μL potassium iodide 1M and incubated for 1 h. Absorbance was read at 390 nm. The standard curve was prepared with different dilutions of a solution containing 5,7 μL 30% hydrogen peroxide in

10 mL of distilled water. H_2O_2 was expressed in μmol for grame of fresh weight (μmol. gFW^{-1}).

The effect of single factors as well as their two-way interactions in leaf H_2O_2 level was determined by ANOVA (F: 2.7, gl: 22, p=0.0005). The analysis showed differences between treatments (F: 2.42, gl: 5, p=0.0416) and sampling times (F: 6.24, gl: 4, p=0.0002), but the interaction between sampling times and treatments was not significant (F: 0.79, gl: 12, p=0.6609).

Level of H_2O_2 in leaf showed a decreasing trend throughout the sampling period (Figure 2), as days 28 and 21 presented significantly lower H_2O_2 (0.48 and 0.51 μmol $H_2O_2.gFW^{-1}$) than the other sampling times (0.96, 0.92, and 0.74 μmol $H_2O_2.gFW^{-1}$ on days 0, 7, and 14, respectively). Significant differences were only observed between treatments at 14 and 28 days. The T1 (1,25 μmol. gFW^{-1}) showed the highest value for 14 days and for 28 days after inoculation, T1 (0.65 μmol. gFW^{-1}) and T5 (0.6 μmol. gFW^{-1}) showed a significantly higher level of H_2O_2 compared with the other treatments and control.

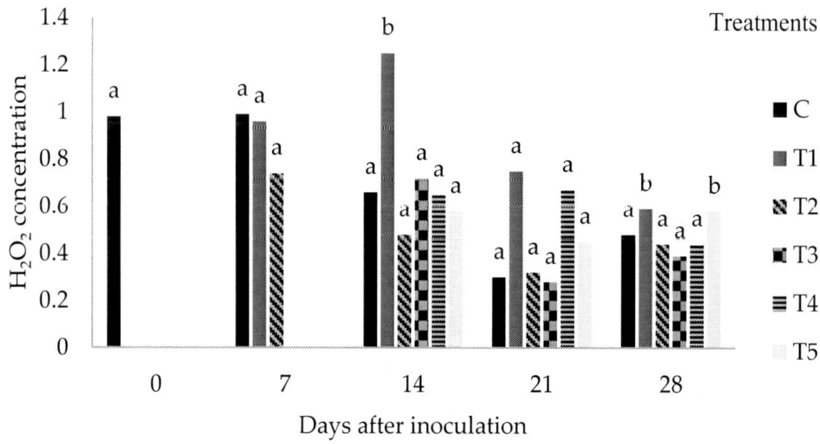

Figure 2. Hydrogen peroxide (H_2O_2) leaf concentration (μmol. gFW^{-1}). Treatment means with the same letter are statistically similar inside each time (p>0.05). Treatments with (*) presented statistically different means (p<0.05). C: control, T1: *B. velezensis* EM-A8 in F1, T2: *B. velezensis* EM-A8 in F2, T3: *B. velezensis* EM-A8 in F1 + *E. turcicum*, T4: *B. velezensis* EM-A8 in F2 + *E. turcicum*, T5: *E. turcicum*.

2.4. HPLC Determination of Salicylic Acid, Abscisic Acid, Indoleacetic Acid, and Jasmonic Acid Leaf Concentration

Times selected for the determination of phytohormones by liquid chromatography were those in which both microorganisms (pathogen and antagonist) coincided (14-, 21- and 28-days' post-inoculation). The levels of salicylic acid (SA), jasmonic acid (JA), indoleacetic acid (IAA), and abscisic acid (ABA) were measured on lyophilized leaves. For the extraction, 0.1 g of dry leaves was homogenized with distilled deionized water with 1% acetic acid. Samples were centrifuged at 5000 rpm for 15 minutes, and the supernatant was added with 1.5 ml of ethyl ether. The mixture was centrifuged again, and then the organic phase was recovered and evaporated. Hormone determination was carried out by HPLC-MC service provided by Grupo de Fisiología Vegetal of Universidad Nacional de Río Cuarto. Analyses were performed using an Alliance 2695 (Separation Module, Waters, USA) quaternary pump equipped with an auto-sampler. A Restek C18 (Restek, USA) column (2.1 x 100 mm, 5 μm) was used at 28°C, with an injected volume of 10 μL. The binary solvent system used for elution gradient consisted of 0.2% acetic acid in H_2O (solvent B) and MeOH (solvent A), at a constant flow rate of 200 μL min^{-1}. A linear gradient profile with the following proportions (v/v) of solvent A was applied [t (min), % A]: (0, 40), (25, 80), with 7 min for re-equilibration. MS/MS experiments were performed on a Micromass Quatro UltimaTM PT double quadrupole mass spectrometer (Micromass, Manchester City, UK). All analyses were performed using a turbo ion spray source (ESI) in negative ion mode with the following settings: capillary voltage -3250 V, energy cone 35 V, RF Lens1 (20), RF Lens2 (0.3), source temperature. 100°C, desolvation temp. 350°C, gas cone 100 L h^{-1}, gas desolvation 700 L h^{-1}, collision cell potential 15 V, and multiplier (650). MS/MS parameters were optimized in infusion experiments using individual standard solutions of each hormone. MS/MS product ions were produced by collision-activated dissociation of selected precursor ions in the collision cell of the double quadrupole mass spectrometer, and mass was analyzed using the second analyzer of the instrument. In negative mode, the spectrum for each hormone gave the deprotonated molecule [M–H]. Quantitation was performed by injection of samples in multiple reaction monitoring (MRM) modes, since many compounds could present the same nominal molecular mass. The combination of parent mass and unique fragment ions was used to selectively monitor hormones. MRM acquisition was performed by monitoring the 263/153 and 269/159 for ABA and (2 H 6)-ABA, 209/53 and 215/53 for JA and (2 H 6)-JA, 137/93 and 141/97 for SA and (2 H 4)-SA, and 174/130 and 179/135 for IAA and (2 H 5)-IAA, with dwell 2000 ms for each transition. Data was acquired and analyzed using MassLynxTM 4.1 and QuanLynxTM 4.1 (Micromass, Manchester, UK) software. For quantification, values were obtained from a calibration curve previously constructed using known amounts of each hormone and their pure standard (Sigma, St. Louis, MO, USA)/deuterated internal standard ratio.

2.4.1. Salicylic Acid Leaf Level

The effect of single factors as well as their two-way interactions in SA concentration was determined by ANOVA (F: 3.57, gl: 18, $p<0.0001$). The analysis showed a significant interaction between sampling times and treatments (F: 3.62, gl: 10, $p=0.0005$). After 14- and 28-days post-inoculation, the control showed significantly higher SA levels compared to the treatments ($p<0.0005$). However, there was no significant difference between treatments (Figure 3). On day 14, the C showed 66.15 nmol SA. gFW-1, while the media for treatment was 26.3 nmol SA. gFW-1. For day 28, C showed 24.95 nmol SA. gFW-1, and the treatments ranged from 19.94 to 20.9 nmol SA. gFW-1.

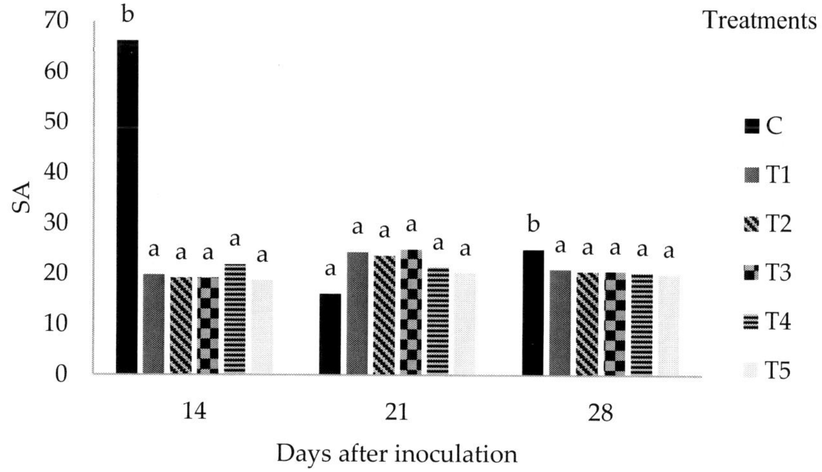

Figure 3. SA leaf concentration determined by liquid chromatography (nmol. gFW^{-1}) means. Treatment means with the same letter are statistically similar inside each time ($p>0.05$). C: control, T1: *B. velezensis* EM-A8 in F1, T2: *B. velezensis* EM-A8 in F2, T3: *B. velezensis* EM-A8 in F1 + *E. turcicum*, T4: *B. velezensis* EM-A8 in F2 + *E. turcicum*, T5: *E. turcicum*.

2.4.2. Abscisic Acid Leaf Level

The effect of single factors as well as their two-way interactions in ABA level was determined by ANOVA (F: 2.17, gl: 18, p=0.0093). The analysis showed differences between sampling times, as day 21 presented significantly lower ABA concentration (1.57 μg. gDM^{-1}) than days 14 and 28 (2.2 and 2.56 μg. gDM^{-1}, respectively). There were no statistical differences between treatments (F: 0.37, gl: 5, p=0.8707) nor the interaction between these two factors (F: 0.7, gl: 10, p=0.7195) (Figure 4).

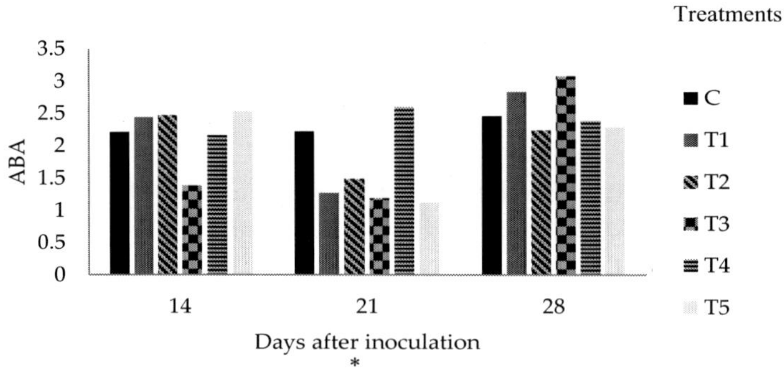

Figure 4. ABA leaf concentration (μg. gDW^{-1}) means. Treatment means with the same letter are statistically similar inside each time (p>0.05). C: control, T1: *B. velezensis* EM-A8 in F1, T2: *B. velezensis* EM-A8 in F2, T3: *B. velezensis* EM-A8 in F1 + *E. turcicum*, T4: *B. velezensis* EM-A8 in F2 + *E. turcicum*, T5: *E. turcicum*.

2.4.3. Indoleacetic Acid Leaf Level

The effect of single factors as well as their two-way interactions in leaf IAA level was evaluated by ANOVA (F: 0.96, gl: 18, p=0.5127). The analysis did not show differences between treatments (F: 0.77, gl: 5, p=0.5737) nor sampling times (F: 0.11, gl: 2, p=0.8979). The interaction between these two factors was not significant (F: 1.27, gl: 10, p=0.2620). Indole acetic acid (IAA) means leaf concentration was 26.21 ± 2.78 μg. gDM^{-1}

2.4.4. Jasmonic Acid Leaf Level

The effect of single factors as well as their two-way interactions was determined by ANOVA (F: 2.93, gl: 18, p=0.0005). Treatments did not affect significantly JA concentration (F:1.04, gl: 5, p=0.3969), but sampling times presented differences (F:10.23, gl: 2, p=0.0001). The interaction between these two factors was not significant (F:0.55, gl: 10, p=0.8491). In this sense, day 21 showed a higher JA concentration (0.94 ng. gDM^{-1}) than days 14 and

28, with 0.33 and 0.49 ng. gDM^{-1}, respectively. On day 28, C showed a significant increase (F: 18.6, gl: 5, p<0.0001) in the level of this compound (1.20 ng. gDM^{-1}), compared to the other treatments, which ranged from 0.33 to 0.55 ng. gDM^{-1} (Figure 5).

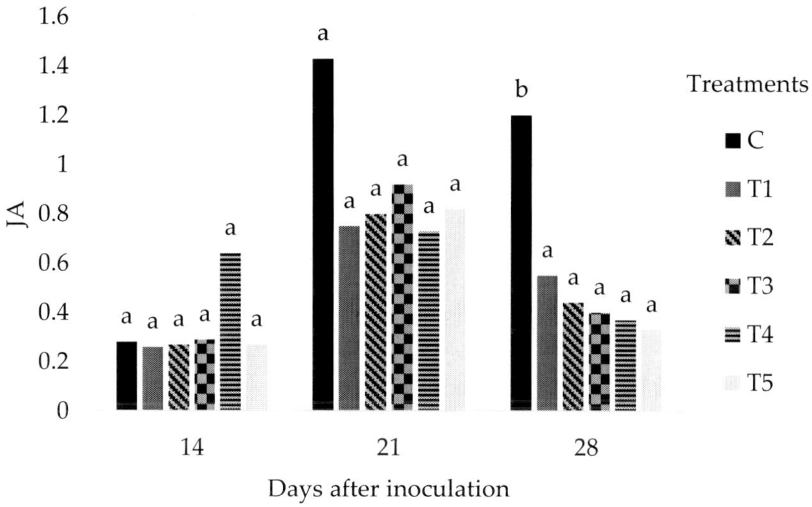

Figure 5. JA (ng.gDW^{-1}) leaf concentration means. Treatment means with the same letter are statistically similar inside each time (p>0.05). C: control, T1: *B. velezensis* EM-A8 in F1, T2: *B. velezensis* EM-A8 in F2, T3: *B. velezensis* EM-A8 in F1 + *E. turcicum*, T4: *B. velezensis* EM-A8 in F2 + *E. turcicum*, T5: *E. turcicum*.

2.5. Phenolic Compounds at the Leaf Level

Total phenolic compounds were determined as described in Nadernejad et al. (2013). 0.1 g of lyophilized leaf tissue was incubated for 2 h with 3 ml of a solution of 80% methanol with 1% hydrochloric acid. Samples were then centrifuged for 5 minutes at 3000 rpm. 0.1 ml of supernatant was mixed with 0.75 ml Folin-Ciocalteu reagent (10%) and let sit for five minutes before adding 0.75 ml Na_2CO_3. Absorbance was read after 90 minutes of reaction at 725 nm. The standard curve was prepared with different concentrations of gallic acid, ranging from 0,5 to 500 mg. L^{-1}. Results were expressed as mg gallic acid equivalent per gram of fresh weight. Results were expressed in mg gallic acid equivalent (GA) per gram of dry weight (mgGA.gDW^{-1}).

The effect of single factors as well as their two-way interactions in leaf phenolic compounds concentration was evaluated by ANOVA (F: 7.15, gl: 22, p<0.0001). The analysis showed differences between sampling times (F: 2.80, gl: 4, p=0.03), but there were no differences found between treatments (F: 1.24, gl: 5, p=0.2971), nor was the interaction between these two factors significant (F: 0.71, gl: 12, p=0.7415). Days 0 and 7 (15.45 and 13.93 mgGA.gDW^{-1}, respectively) showed significantly higher average phenolic compounds concentration than the other sampling times (14, 21, and 28 days), which ranged from 11.69 to 12.64 mgGA.g DW $^{-1}$, on average.

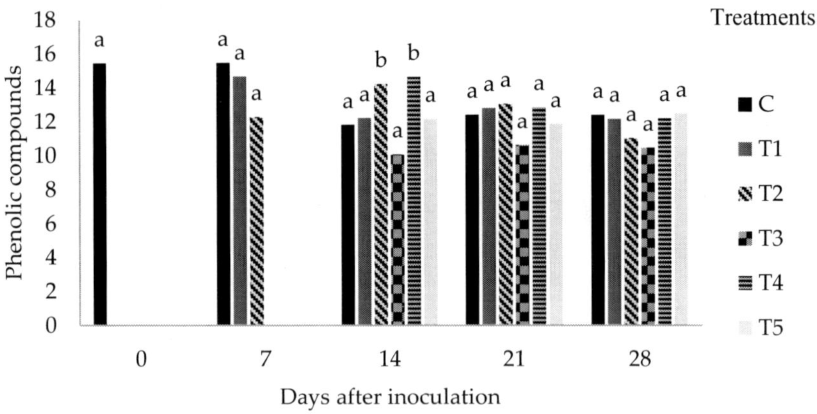

Figure 6. Phenolic compounds leaf concentration (mg GA equivalent.g DW^{-1}). Treatment means with the same letter are statistically similar inside each time (p>0.05). C: control, T1: *B. velezensis* EM-A8 in F1, T2: *B. velezensis* EM-A8 in F2, T3: *B. velezensis* EM-A8 in F1 + *E. turcicum*, T4: *B. velezensis* EM-A8 in F2 + *E. turcicum*, T5: *E. turcicum*.

On day 14, treatments revealed statistical differences (F: 2.82, gl: 5, p=0.0398). T2 and T4 presented higher phenolic compound concentration (14.23 and 15.2 mgGA.g DW $^{-1}$, respectively) than T3 (10.08 mgGA.g DW $^{-1}$), C (11.82 mgGA.g DW $^{-1}$), T5 (12.18 mgGA.g DW $^{-1}$), and T1 (12.22 mgGA.g DW $^{-1}$) (Figure 6).

2.6. Principal Component Analysis

Analysis of principal components explained 75.6% of the total variation observed. Leaf cell membrane stability correlated negatively with SA, AIA, and JA leaf concentration, whereas phenolic compounds, ABA, and H_2O_2 leaf level correlated positively. SA correlated positively with JA. The control was associated with the variables JA and SA, while the other treatments, mainly T2, were associated with membrane stability (Figure 7).

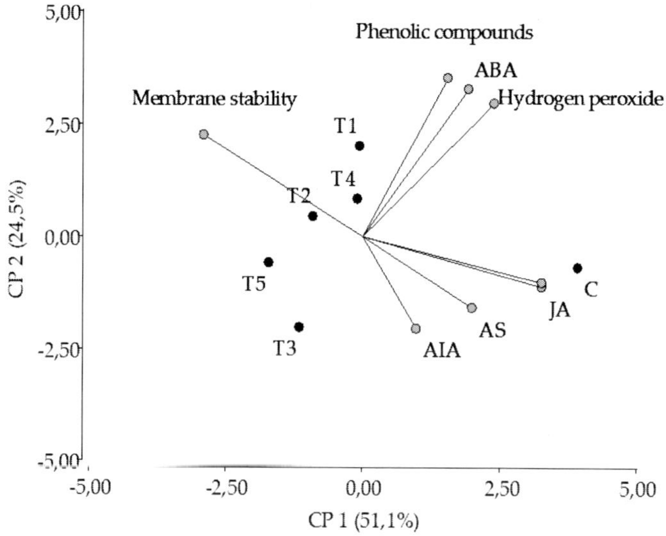

Figure 7. Principal component analysis including Membrane stability, Phenolic compounds, H_2O_2, and phytohormones determinations (SA, JA, AIA, ABA).

2.7. Characterization of NCLB-Causing Pathogenic Agent Isolates

Pathogenicity and molecular identification of three randomly selected isolates were made. Isolated were collected from NCLB lesions of maize leaves from the south Córdoba province (Argentina).

2.7.1. Pathogenicity Assay of the Isolates

Two hybrids (I797 and 7123PW) with different NCLB resistance were sown in 10L pots under greenhouse conditions in a randomized complete blocks design with three replicates. At four leaves stage, plants were sprayed with 10 mL of a conidial solution prepared with sterile water and Tween 20 (1%) and conidia scraped from potato dextrose-agar cultures of the collected isolates. After inoculation, pots were covered with polyethylene for 24 h to reach maximum relative humidity. Severity, according to Bleicher (1988), and the number of affected leaves were assessed every week for 35 days. Selected isolates were the numbers 5, 17, and 30, and the results found are summarized in Table 1. They indicate that there is variability in the isolates' pathogenicity, as isolate 30 failed at causing disease symptoms in the moderately resistant hybrid, but isolates 5 and 17 succeeded. On the other hand, the moderately susceptible hybrid showed typical NCLB lesions after inoculation with the three isolates.

Table 1. Average number of affected leaves and severity of NCLB measured on a moderately resistant hybrid and on a moderately susceptible hybrid after inoculation with isolates 5, 17, and 30

	N° affected leaves			Severity (%)		
Isolates	5	17	30	5	17	30
Moderately resistant hybrid	0,33	1,67	0,00	0,33	0,60	0,00
Moderately susceptible hybrid	3,93	3,47	3,93	1,40	1,33	1,33

2.7.2. Molecular Identification of the Three Selected Isolated

The genomic DNA (DNAg) extraction was performed over conidia scraped from potato-dextrose agar Petri cultures. Conidia were frozen, pulverized, and added with 700 µL of CTAB buffer at 65°C and 7 µL 2-mercaptoethanol. Tubes containing this mixture were incubated at 65°C for 30 minutes. After that, 300 µL of a mixture of chloroform: isoamyl alcohol (24:1 v/v) was added, and the tubes were centrifuged for 5 minutes at 12000 RPM. The aqueous phase was recovered and transferred to new tubes, and then added with 600 µL isopropanol. The new tubes were then centrifuged at 10000 RPM for 5 minutes, and the supernatant was discarded. The pellets were resuspended with 200 µL nuclease-free water and kept under -20°C. The presence of DNAg was confirmed by an electrophoretic run.

DNAg sequencing of ITS1 and ITS4 was performed by Macrogen (the Republic). The consensus sequence was created using the software BioEdit. BLAST queries were performed to compare the consensus sequences with those of other fungi deposited in the GenBank database (https://blast.ncbi.nlm.nih.gov/Blast.cgi). Genomic DNA of isolates 5, 17, and 30 was extracted from pure cultures and sequenced. The analysis of the consensus sequences of the isolates 5, 17, and 30 showed 100%, 99.82% and 99.82% respectively, of similitude with *Exserohilum turcicum* strains deposited on GenBank. The pairings are shown in Table 2.

Table 2. Taxon association of the isolates 5, 17, and 30 through molecular analysis of the ITS sequence

Isolate	Taxon pairing	Maximum score	Query coverage	% identification	Accession (GenBank)
5	*Exserohilum turcicum*	1029	97%	100	MG727608.1
17	*Exserohilum turcicum*	1022	97%	99.82	NR_163537.1
30	*Exserohilum turcicum*	1027	87%	99.82	PQ289223.1

3. Discussion

Corn is a cereal of great economic importance worldwide, whether as human food (one of the oldest known food grains), as livestock feed, or as a source of a large number of industrial products (Golik et al., 2018). However, its production can be reduced by several diseases such as NCLB. For this reason, in this work, maize plants were treated with two different formulations of *B. velezensis* EM-A8, a biocontrol agent of *E. turcicum*, the etiological agent of NCLB. Different parameters and biochemical compounds involved in the defensive response were measured.

In our work, we can confirm that the studied BCA is able to colonize the phyllosphere because the survival of *B. velezensis* EM-A8, assessed through colony-forming units, was in the order of 5 logarithms for both F1 and F2 after a week of their application. This indicates a proper survival of the bacteria on the maize leaves, an expected attribute for native foliar bacteria (Sartori et al., 2015, 2016, 2017a).

On the other hand, we evaluated membrane stability, which is a strong indicator of pathogen spread and can be measured by electrolyte leakage (Bach-Pages and Preston, 2018). In the greenhouse trials, the application of the biocontroller presented significantly higher membrane stability (%MS) compared to the control, although there were no differences compared to the pathogen-inoculated plants. Besides, %MS was similar among treatments, except for days 7 and 21. In the first case, cell membranes in T2 were 46% more stable than C and 63% more stable than T1. After 21 days of inoculation, T1 was up to 84% superior to the other treatments. These results indicate that the pathogen did not significantly affect the membrane integrity, nor did the plant experience a hypersensitive response during the sampling periods analyzed. Also, the BCA did not stress the plant, as the membrane stability remained constant in *B. velezensis* EM-A8 treated plants. Our results differ from those obtained by Mulk et al. (2022), who found higher electrolyte leakage in control wheat plants than in *B. velezensis-inoculated* plants. However, Omara & Abdelaal (2018) found that biocontrol agents, particularly *B. subtilis* and *Trichoderma viride,* decreased the final disease severity and electrolyte leakage on wheat plants under leaf rust stress. On the other hand, our results do not agree with the results obtained by Silveira et al. (2019), who found greater electrolyte leakage associated with less membrane integrity in maize leaves infected with *E. turcicum* after 10 days after inoculation. However, recovery from electrolyte leakage during the infection period was observed by Meena et al. (2021) in banded leaf and sheath blight-resistant maize inbreds, indicating the importance of the host defense mechanisms during disease development. These results demonstrate that the recovery of electrolyte leakage is a desirable trait and that membrane stability is host-dependent, pathogen-dependent, and the environment in which they develop.

With respect to H_2O_2, in our study, after 14 days pos inoculation, T1 showed 89% more than C, and at 28 days, the pathogen-inoculated plants presented 20% more of this compound concentration than C. These results agree with Silveira et al. (2019), who measured up to a 25% increase in H_2O_2 level in *E. turcicum* inoculated plants, compared to healthy ones. These differences between treatments could be explained by the pathogen´s hemibiotrophic lifestyle. In this sense, when *E. turcicum* starts colonizing, hyphal growth can be restrained because of reactive oxygen species (ROS) accumulation, such as H_2O_2, at the infection site, a reaction called hypersensitive response that leads to programmed cell death (Apel & Hirt, 2004). It is known that H_2O_2 accumulates as a normal result of oxidative metabolism and after biotic and abiotic stresses. Besides, it participates in the hypersensitive response by inducing cell death and creating a physical isolation that avoids pathogens′ spread (Liu et al., 2007). In this sense, in our results, the increase in H_2O_2 observed on day 14 in T1 could be related to the hypersensitive response. Regarding the biocontroller influence over H_2O_2 levels, *B. velezensis* EM-A8 showed a differential effect depending on the formulation. Whereas T1 showed higher levels of this compound on days 14 and 28 compared to healthy plants, T2, T3, and T4 triggered similar H_2O_2 concentrations to C. In the literature reviewed, there is no clear trend regarding what happens with this compound. Some authors determined a decrease in peroxides in plants when treated with different biocontrol agents (Azeem et al., 2022, Kazerooni et al., 2021), while others determined increases in the levels of this compound (Chen et al., 2021). This means that it depends on the pathosystem being evaluated and the growing conditions.

Regarding AS, it is known that it participates in several physiological processes, including resistance against biotrophic pathogens. In this sense, SA can activate the expression of disease-resistance genes, regulate cell wall strengthening, enhance resistance to oxidative stress, and even trigger programmed cell death (Klessing & Choi, 2018, Koo et al., 2020, Liu et al., 2007 & Pal & Gardener, 2006). Moreover, SA can increase chlorophyll content, thus enhancing crop yields (Song et al., 2023). In our study, significant differences were observed between the levels of this compound at 14 days post-inoculation, as the C showed levels 3.75 times higher than treatment, which had an average of 20 nmol SA.gFW^{-1}. SA is the main signaling molecule for Systemic Acquired Resistance (SAR), which provides resistance against biotrophic pathogens. However, the SA leaf concentration in our study was higher in C than in the other treatments. These are in contrast with Morris et al. (1998) findings in *Colletotrichum graminicola* and *Bipolaris maydis* infected maize leaves. These authors found increased SA levels in infected maize leaves. These contrasts in the bibliography could also be explained by the *E. turcicum* hemibiotrophic lifestyle. During a biotrophic pathogen disease development, SA levels are strongly correlated with disease resistance. On the contrary, SA's role during necrotrophic and hemibiotrophic pathogen infection is less predictable [40]. On the other hand, the biocontroller's effect on SA concentration depends on the pathosystem analyzed. In this sense, Agostini et al. (2019) analyzed hormonal changes after treatment with Trichoderma atroviride and *Fusarium verticillioides,* and they found the most significant changes in the SA levels. In this case, *F. verticillioides* also acts as a biotrophic pathogen at the early stages of the disease, coinciding with an increase in SA levels. In our study, the treatment with the biocontroller was statistically similar to T5, plants with *E. turcicum.* These results could suggest a priming effect of the biocontroller for the disease.

Phytohormonal crosstalk adjusts the balance between growth and defense by regulating resource distribution in order to optimize the plant's fitness to changing environments. Abscisic acid (ABA) is involved in developmental processes as well as in growth regulation during abiotic stresses. More recently, ABA has been linked to defense response (Denancé et al., 2013). In this context, as in most phytohormones, ABA can function as a positive or negative regulator of the plant defense depending on the pathosystem analyzed. However, in pathosystems that involve hemibiotrophic pathogens, ABA levels generally remain considerably constant. For example, in maize plants infected with the hemibiotrophic pathogen *F. verticillioides* along with the biocontroller *T. atroviride*, Agostini et al. (2019) found that there were no significant changes in ABA levels among treatments. On the other hand, auxin is a growth-promoting phytohormone. Indoleacetic acid (IAA) is the main auxin found in plants. It is reported to generally impair immunity because it is a SA signaling pathway antagonist (Ye et al., 2019), and it is known that IAA concentration decreases during hypersensitive response (Yousaf et al., 2022). However, there are some exceptions. For example, Degani *et al.* (Degani et al., 2015) found that IAA in culture media suppressed *Haropophora maydis* growth. In another study, Ye et al. (2019) proved that a quantitative disease-resistance locus that protects against *Gibberella* stalk rot promotes the biosynthesis of IAA. ABA and AIA levels remained constant between sampling times in our greenhouse trials. Also, there were no differences in these phytohormone concentrations that could be associated with the application of the biocontroller or the pathogen.

On the opposite side of SA, JA concentration is associated with protection against necrotrophic pathogens and Induced Systemic Resistance (ISR) activation. There is an antagonism between ISR and SAR, since they provide protection against different kinds of pathogens, although synergism can occur in some cases (McLaughlin et al., 2023). However, in our investigation, we could not discriminate between the combined treatments and the BCA treatments in terms of JA concentration. Yet, the control showed significantly higher concentration compared to the other treatments. Regarding JA's role during NCLB disease, Lipps et al. (2023) established that several pathways involved in JA biosynthesis were associated with NCLB resistance. This contrasts with what Agostini et al. (2019) found in maize silks. These authors analyzed hormonal changes after treatment with *T. atroviride* and *F. verticillioides* and discovered that *F. verticillioides* treatment, independently of the *Trichoderma* presence, caused higher JA levels.

Phenolic compounds play an important role during biotic and abiotic stresses. Polyphenols are secondary metabolites of importance in extreme survival conditions. Besides, these compounds act as antioxidants by protecting cell membranes from oxidative stress caused by reactive oxygen species, provide defense against pathogens and herbivory, and enhance water stress tolerance. Particularly, phenolic compounds participate in cell wall strengthening (Kulbat, 2016), rendering the plant more resistant to fungal infection. In our study, on day 14, treatments that included F2 (T2 and T4) showed a significant increase in phenolic compounds concentration. This could indicate an enhanced defensive response after F2 application. In this sense, a higher phenolic compound concentration enhanced maize resistance to pathogens such as *F. verticillioides* and southern corn leaf blight (Bernardi et al., 2018 & Vasmatkar et al., 2019). In a meta-analysis, Wallis and Galarneau (2020) proved that pathogenic fungi, mostly necrotrophic, failed to induce phenolic compound concentration, in contrast to beneficial fungi and both pathogenic and beneficial bacteria. In this context, our study showed that infected plants presented similar phenolic compound concentrations to healthy ones. Regarding BCA influence over phenolic compounds concentration, Li et al. (2021) analyzed the expression of maize resistance-related genes induced by the biocontrol bacteria *Xenorhabdus budapestensis* cell-free culture media. They found that the expression of the PAL gene, which is crucial in the phenylpropanoid metabolism, increased from 12 to 48 hours after the treatment.

Concerning the principal component analysis, SA correlated negatively with membrane stability, an expected result as SA induces cell death during the hypersensitive response (Koo et al., 2020). However, in our study, SA did not correlate with hydrogen peroxide concentration, suggesting that cell death could be achieved by other kinds of reactive oxygen species or that antioxidant systems are actively participating in detoxifying these compounds. In this sense, there is evidence that SA can also enhance the activities and expression of antioxidant enzymes, including peroxidase (Ali et al., 2023). Contrary to what was expected, SA correlated positively with JA. In most cases, there is an antagonism between ISR and SAR, since they provide protection against different kinds of pathogens, although there is evidence that synergism can occur in some cases (McLaughlin et al., 2023).

Finally, and in order to identify the pathogen isolates used in the greenhouse experiments, a pathogenicity assay was carried out with 3 isolates (5, 17, and 30) collected from the South of Córdoba province. Results showed variability among the *E. turcicum* strains. While isolate 17 affected a larger number of leaves in the moderately resistant hybrid than isolate 5, isolate 30 failed at producing disease symptoms. However, on the moderately susceptible hybrid, the three isolates caused lesions to a greater extent compared to the moderately resistant hybrid. Physiological races of the pathogen are determined according to maize qualitative resistance *Ht* genes. For example, race 0 is avirulent in plants carrying *Ht* resistance genes, while race 1 is avirulent to maize carrying *Ht1* genes. The differences found between strains in the pathogenicity trial indicate there is genetic variation regarding the pathogen races. In this sense, Navarro et al. (2021b) found that the pathogen race 0 was the most frequent in Argentina (83%), followed by race

1 and race 23N. Despite the variability in *E. turcicum* races, the presence of *Ht* genes in the most cultivated hybrids in Argentina is not generally provided.

For further characterization, the three isolates were identified through molecular methods. For this, genomic DNA was extracted, and ITS1 and ITS4 were sequenced to compare with other fungal segments deposited in the GeneBank database. In our identification of fungal isolates from maize NCLB lesions, we found that the three isolates corresponded to *E. turcicum*. In a previous study, Hellrigl (2022) identified ten isolates from *Sorghum halepense* leaves with NCLB symptoms, found in maize crops in the South of Córdoba province. This author found that five of them corresponded to *E. turcicum* strains. However, it may be necessary to perform a multilocus analysis to determine the species of *Exserohilum* among the genre.

Results obtained in this research allow us to learn more about the maize pathogen *E. turcicum* and its behavior in the presence of the BCA, *B. velezensis* EM-A8. These results provide the opportunity to continue advancing our knowledge and applications of BCA in field trials where environmental conditions are less controlled and the compounds evaluated may behave differently.

References

Agostini, R. B., Postigo, A., Rius, S. P., Rech, G. E., Campos-Bermudez, V. A. & Vargas, W. A. (2019) Long-Lasting Primed State in Maize Plants: Salicylic Acid and Steroid Signaling Pathways as Key Players in the Early Activation of Immune Responses in Silks. *Mol. Plant-Microbe Interact.* 32, 90-106, doi:10.1094/MPMI-07-18-0208-R.

Ali, Q., Ahmad, M., Kamran, M., Ashraf, S., Shabaan, M., Babar, B. H., Zulfiqar, U., Haider, F. U., Ali, M. A., & Elshikh, M. S. (2023) Synergistic Effects of Rhizobacteria and Salicylic Acid on Maize Salt-Stress Tolerance. *Plants,* 12, doi:10.3390/plants12132519.

Apel, K. & Hirt, H. (2004) Reactive Oxygen Species: Metabolism, Oxidative Stress, and Signal Transduction. *Annu. Rev. Plant Biol.* 55, 373–399, doi:10.1146/annurev.arplant.55.031903.141701.

Azeem, M., Haider, M. Z., Javed, S., Saleem, M. H. & Alatawi, A. (2022) Drought Stress Amelioration in Maize (*Zea mays* L.) by Inoculation of *Bacillus* Spp. Strains under Sterile Soil Conditions. *Agric.* 12, doi:10.3390/agriculture12010050.

Bach-Pages, M. & Preston, G. M. (2018) Methods to Quantify Biotic-Induced Stress in Plants. *Methods Mol. Biol.* 1734, 241-255, doi:10.1007/978-1-4939-7604-1_19.

Bernardi, J., Stagnati, L., Lucini, L., Rocchetti, G., Lanubile, A., Cortellini, C., De Poli, G., Busconi, M. & Marocco, A. (2018) Phenolic Profile and Susceptibility to Fusarium Infection of Pigmented Maize Cultivars. *Front. Plant Sci.* 9, 1-13, doi:10.3389/fpls.2018.01189.

Bleicher, J. (1988). Níveis de resistência a *Helminthosporium turcicum* Pass. Em três ciclos de seleção em milho pipoca (*Zea mays* L.). *Piracicaba.* Tese (Doutorado) - ESALQ – SP, Brasil. 130p.

Bolsa de Comercio de Rosario, (BCR) (2025). https://www.bcr.com.ar/es/mercados/gea/estimaciones-nacionales-de-produccion/estimaciones.

Chaudhry, V., Runge, P., Sengupta, P., Doehlemann, G., Parker, J. E. & Kemen, E. (2021) Shaping the Leaf Microbiota: Plant-Microbe-Microbe Interactions. *J. Exp. Bot.,* 72, 36-56, doi:10.1093/jxb/eraa417.

Chen, Z., Zhao, L., Dong, Y., Chen, W., Li, C., Gao, X., Chen, R., Li, L. & Xu, Z. (2021) The Antagonistic Mechanism of Bacillus Velezensis ZW10 against Rice Blast Disease: Evaluation of ZW10 as a Potential Biopesticide. *PLoS One,* 16, 1-14, doi:10.1371/journal.pone.0256807.

Degani, O., Drori, R. & Goldblat, Y. (2015) Plant Growth Hormones Suppress the Development of Harpophora Maydis, the Cause of Late Wilt in Maize. *Physiol. Mol. Biol. Plants,* 21, doi:10.1007/s12298-014-0265-z.

Denancé, N., Sánchez-Vallet, A., Goffner, D. & Molina, A. (2013) Disease Resistance or Growth: The Role of Plant Hormones in Balancing Immune Responses and Fitness Costs. *Front. Plant Sci.* 4, 1-12, doi:10.3389/fpls.2013.00155.

De Rossi, R. L., Guerra, F. A., Plazas, M. C., Vuletic, E. E., Brücher, E., Guerra, G. D. & Reis, E. M. (2022) Crop Damage, Economic Losses, and the Economic Damage Threshold for Northern Corn Leaf Blight. *Crop Prot.* 154, 105901, doi:10.1016/j.cropro.2021.105901.

Formento, N. A. (2010) Enfermedades Foliares Reemergentes Del Cultivo de Maíz: (*Puccinia sorghi* y *Puccinia polysora*), Tizón Foliar (*Exserohilum turcicum*) y Mancha Ocular (*Kabatiella zeae*). *Estac. Exp. Agropecu.* Paraná, 89-100.

Galiano-Carneiro, A. L. & Miedaner, T. (2017) Genetics of Resistance and Pathogenicity in the Maize/Setosphaeria Turcica Pathosystem and Implications for Breeding. *Front. Plant Sci.* 8.

Golik, S. I., Larran, S., Gerard, G. S. & Fleitas, M. C. (2018). Maíz, importancia, origen, sistemática, morfología y composición química. In: *Cereales de Veran.* Editorial: Universidad Nacional de La Plata. Facultad de Ciencias Agrarias y Forestales. ISBN: 978-950-34-1658-7.

Hellrigl, J. (2022) Tesis de grado: Evaluación de la maleza sorgo de Alepo como reservorio del agente causal del tizón foliar común (*Exserohilum turcicum*) en maíz.

Kazerooni, E. A., Maharachchikumbura, S. S. N., Al-Sadi, A. M., Kang, S. M., Yun, B. W. & Lee, I. J. (2021) Biocontrol Potential of *Bacillus amyloliquefaciens* against *Botrytis pelargonii* and *Alternaria alternata* on *Capsicum annuum. J. Fungi* 7, doi:10.3390/jof7060472.

Klessig, D. F., Choi, H. W. & Dempsey, D. A. (2018) Systemic Acquired Resistance and Salicylic Acid: Past, Present, and Future. *Mol. Plant-Microbe Interact.* 31, 871-888, doi:10.1094/MPMI-03-18-0067-CR.

Koo, Y. M., Heo, A. Y. & Choi, H. W. (2020) Salicylic Acid as a Safe Plant Protector and Growth Regulator. *Plant Pathol. J.* 36, 1-10, doi:10.5423/PPJ.RW.12.2019.0295.

Kotze, R. G., van der Merwe, C. F., Crampton, B. G. & Kritzinger, Q. (2019) A Histological Assessment of the Infection Strategy of *Exserohilum turcicum* in Maize. *Plant Pathol.* 68, 504-512, doi:10.1111/ppa.12961.

Kulbat, K. (2016) Biotechnology and Food Sciences The Role of Phenolic Compounds in Plant Resistance. *Biotechnol Food Sci.* 80, 97-108.

Legein, M., Smets, W., Vandenheuvel, D., Eilers, T., Muyshondt, B., Prinsen, E., Samson, R. & Lebeer, S. (2020) Modes of Action of Microbial Biocontrol in the Phyllosphere. *Front. Microbiol.* 2020.

Li, B., Kong, L., Qiu, D., Francis, F. & Wang, S. (2021) Biocontrol Potential and Mode of Action of Entomopathogenic Bacteria *Xenorhabdus budapestensis* C72 against *Bipolaris maydis*. *Biol. Control.* 158, doi:10.1016/j.biocontrol.2021.104605.

Lipps, S., Lipka, A. E., Mideros, S. & Jamann, T. (2023). Inhibition of Ethylene Involved in Resistance to *E. Turcicum* in an Exotic-Derived Double Haploid Maize Population. *Front. Plant Sci. 2023*, 14, 1-16, doi:10.3389/fpls.2023.1272951.

Liu, Y., Ren, D., Pike, S., Pallardy, S., Gassmann, W. & Zhang, S. (2007) Chloroplast-Generated Reactive Oxygen Species Are Involved in Hypersensitive Response-like Cell Death Mediated by a Mitogen-Activated Protein Kinase Cascade. *Plant J.* 51, 941-954, doi:10.1111/j.1365-313X.2007.03191.x.

McLaughlin, M. S., Roy, M., Abbasi, P. A., Carisse, O., Yurgel, S. N., & Ali, S. (2023) Why Do We Need Alternative Methods for Fungal Disease Management in Plants? *Plants,* 12, 1-30, doi:10.3390/plants12223822.

Meena, B. R., Yerasu, S. R., Gupta, N. & Singh, J. (2021) Resistance Assessment and Biochemical Responses of Maize Genotypes against *Rhizoctonia Solani* f. Sp. Sasakii Exner Causing Banded Leaf and Sheath Blight Disease. Australas. *Plant Pathol.* 50, 41-49, doi:10.1007/s13313-020-00760-2.

Montemarani, A., Sartori, M., Nesci, A., Etcheverry, M. & Barros, G. (2018) Influence of Crop Residues, Matric Potential and Temperature on Growth of *Exserohilum turcicum* an Emerging Maize Pathogen in Argentina. *Lett. Appl. Microbiol.* 67, 614–619, doi:10.1111/lam.13076.

Morris, S. W., Vernooij, B., Titatarn, S., Starrett, M., Thomas, S., Wiltse, C. C., Frederiksen, R. A., Bhandhufalck, A., Hulbert, S. & Uknes, S. (1998) Induced Resistance Responses in Maize. *Mol. Plant-Microbe Interact.* 11, 643-658, doi:10.1094/MPMI.1998.11.7.643.

Mulk, S., Wahab, A., Yasmin, H., Mumtaz, S., El-Serehy, H. A., Khan, N. & Hassan, M. N. (2022). Prevalence of Wheat Associated *Bacillus* Spp. and Their Bio-Control Efficacy Against *Fusarium* Root Rot. *Front. Microbiol.* 12, 1-13, doi:10.3389/fmicb.2021.798619.

Nadernejad, N., Ahmadimoghadam, A., Hossyinifard, J. & Poorseyedi, S. (2013) Effect of Different Rootstocks on PAL Activity and Phenolic Compounds in Flowers, Leaves, Hulls and Kernels of Three Pistachio (*Pistacia Vera* L.) Cultivars. *Trees - Struct. Funct.* 27, 1681–1689, doi:10.1007/s00468-013-0915-8.

Navarro, B. L., Campos, R. de A., de Godoy Gasparoto, M. C., & Tiedemann, A. von. (2021a) *In vitro and in Planta* Studies on Temperature Adaptation of *Exserohilum turcicum* Isolates from Maize in Europe and South America. *Pathogens* 2021a, 10, 1-16.

Navarro, B. L., Ramos Romero, L., Kistner, M. B., Iglesias, J., & von Tiedemann, A. (2021b) Assessment of Physiological Races of *Exserohilum turcicum* Isolates from Maize in Argentina and Brazil. *Trop. Plant Pathol.*, 46, 371-380, doi:10.1007/s40858-020-00417-x.

Ohm, R. A., Feau, N., Henrissat, B., Schoch, C. L., Horwitz, B. A., Barry, K. W., Condon, B. J., Copeland, A. C., Dhillon, B., Glaser, F., et al. (2012) Diverse Lifestyles and Strategies of Plant Pathogenesis Encoded in the Genomes of Eighteen Dothideomycetes Fungi. *PLoS Pathog.* 8, doi:10.1371/journal.ppat.1003037.

Omara, R. I. & Abdelaal, K. A. A. (2018) Biochemical, Histopathological and Genetic Analysis Associated with Leaf Rust Infection in Wheat Plants (*Triticum Aestivum* L.). *Physiol. Mol. Plant Pathol.* 104, 48-57, doi:10.1016/j.pmpp.2018.09.004.

Pal, K. K. & Gardener, B. M. (2006) Biological Control of Plant Pathogens. Curr. Opin. Biotechnol. 7, 343-347, doi:10.1016/S0958-1669(96)80042-5.

Sartori, M., Nesci, A., Formento, Á. & Etcheverry, M. (2015) Selección de Microorganismos Epifíticos de Maíz Como Potenciales Agentes Debiocontrolde *Exserohilum turcicum. Rev. Argent. Microbiol.* 47, 62-71, doi:10.1016/j.ram.2015.01.002.

Sartori, M., Nesci, A., Montemarani, A., Barros, G., García, J. & Etcheverry, M. (2017a) Preliminary Evaluation of Biocontrol Agents against Maize Pathogens *Exserohilum turcicum* and *Puccinia sorghi* in Field Assays. *Agric. Sci.* 08, 1003-1013, doi:10.4236/as.2017.89073.

Sartori, M., Nesci, A., García, J., Passone, M. A., Montemarani, A. & Etcheverry, M. (2017b) Efficacy of Epiphytic Bacteria to Prevent Northern Leaf Blight Caused by *Exserohilum turcicum* in Maize. *Rev. Argent. Microbiol.* 49, 75–82, doi:10.1016/j.ram.2016.09.008.

Savary, S., Willocquet, L., Pethybridge, S. J., Esker, P., McRoberts, N., & Nelson, A. (2019) The Global Burden of Pathogens and Pests on Major Food Crops. *Nat. Ecol. Evol.* 3, 430-439, doi:10.1038/s41559-018-0793-y.

Silveira, P. R., Milagres, P. O., Corrêa, E. F., Aucique-Pérez, C. E., Wordell Filho, J. A. & Rodrigues, F. A. (2019) Changes in Leaf Gas Exchange, Chlorophyll a Fluorescence, and Antioxidants in Maize Leaves Infected by Exserohilum Turcicum. *Biol. Plant.* 63, 643-653, doi:10.32615/bp.2019.130.

Song, W., Shao, H., Zheng, A., Zhao, L. & Xu, Y. (2023) Advances in Roles of Salicylic Acid in Plant Tolerance Responses to Biotic and Abiotic Stresses. *Plants* 12, 1-23, doi:10.3390/plants12193475.

Vasmatkar, P., Kaur, K., Pannu, P. P. S., Kaur, G. & Kaur, H. (2019) Unraveling the Metabolite Signatures of Maize Genotypes Showing Differential Response towards Southern Corn Leaf Blight by 1H-NMR and FTIR Spectroscopy. *Physiol. Mol. Plant Pathol.* 108, 101441, doi:10.1016/j.pmpp.2019.101441.

Wang, H., Li, H., Zhu, Z. & Wang, X. (2010) Expression of Ht2-Related Genes in Response to the HT-Toxin of *Exserohilum Turcicum* in Maize. *Ann. Appl. Biol.* 156, 111-120, doi:10.1111/j.1744-7348.2009.00369.x.

Wallis, C. M. & Galarneau, E. R. A. (2020) Phenolic Compound Induction in Plant-Microbe and Plant-Insect Interactions: A Meta-Analysis. *Front. Plant Sci.* 11, 1-13, doi:10.3389/fpls.2020.580753.

Welz, H. G. & Geiger, H. H. (2000) Genes for Resistance to Northern Corn Leaf Blight in Diverse Maize Populations. *Plant Breed.* 119, 1-14.

Ye, J., Zhong, T., Zhang, D., Ma, C., Wang, L., Yao, L., Zhang, Q., Zhu, M. & Xu, M. (2019) The Auxin-Regulated Protein ZmAuxRP1 Coordinates the Balance between Root Growth and Stalk Rot Disease Resistance in Maize. *Mol. Plant.* 12, 360-373, doi:10.1016/j.molp.2018.10.005.

Yousaf, M. J., Hussain, A., Hamayun, M., Iqbal, A., Lee, I. J. (2022) Gibberellins Hypersensitivity Hinder the Interaction of Bipolaris Sorokiniana (Scc.) under Cross Talks with IAA and Transzeatin. *J. Plant Interact.* 17, 152-167, doi:10.1080/17429145.2021.2018060.

Zuza, M., Kearney, M., Costa, M., Peralta, V., Salomon, A., & Canale, A. (2022) Evaluación de Híbridos de Maíz Frente a Enfermedades Fúngicas, En El Centro-Sur de La Provincia de Córdoba. *Rev. Ab Intus FAV_UNRC,* 9, 25-34.

Chapter 2

Studies on the Influence and Survival of *Bacillus velezensis* EM-A8, an Antagonist of *Exserohilum turcicum*, in the Maize Phyllosphere

Melina Sartori[1,*]
Daiana García[1]
María Fiamma Grossi Vanacore[1]
Martín Bonacci[1]
Julieta Orlando[2]
Germán Barros[1]
and Andrea Nesci[1]

[1]Laboratorio de Ecología Microbiana, Instituto ICBIA (CONICET-UNRC), Departamento de Microbiología e Inmunología, Facultad de Ciencias Exactas Físico Químicas y Naturales, Universidad Nacional de Río Cuarto, Río Cuarto, Córdoba, Argentina
[2]Laboratorio de Ecología Microbiana, Departamento Ciencias Ecologicas, Facultad de Ciencias, Universidad de Chile, Ñuñoa, Santiago, Chile

Abstract

Biological control is an effective and sustainable alternative to the use of conventional pesticides for plant disease management. *Bacillus velezensis* EM-A8 is a native bacterium of maize leaf phyllosphere antagonist to *Exserohilum turcicum*, a filamentous fungus that causes

* Corresponding Author's Email: msartori@exa.unrc.edu.ar

In: Biocontrol Agents (BCAs)
Editor: Daiana García
ISBN: 979-8-90134-102-5

Northern Corn Leaf Blight (NCLB) disease. The aims of this study were: i) to detect changes that occur in the microbial community in maize phyllosphere after foliar application of *B. velezensis,* and ii) to study the survival of the bacteria in the field. The physiological profile of the microbial communities studied in the maize phyllosphere was determined using a pattern of 31 carbon sources on Biolog EcoplateTM. For this, the following treatments in maize plants were made: *in vitro* (T1- *B. velezensis* EM-A8, T2- *E. turcicum*), field maize trial (T3- Control, T4- *B. velezensis* EM-A8 formulated, T5- *B. velezensis* EM-A8 CN), and maize in growth chamber (T6- Control, T7- *B. velezensis* EM-A8 CN, T8- *E. turcicum*, T9- *B. velezensis* EM-A8 CN + *E. turcicum*). From the absorbance at 160 h of incubation, a dendrogram and profiles were established. Besides, the monitoring of bacteria was made by analyzing samples at different times of maize plants inoculated with the antagonist. For the T3, T4, and T5 (field trial) showed the greatest capacity to utilize carbon sources, and the presence of the antagonist did not alter the metabolic pattern. However, T2 and T6 showed the lowest capacity to use carbon sources, indicating a low physiological diversity. T1, T7, T8, and T9 showed a similar physiological pattern, highlighting the versatility of the antagonist in the use of different substrates. Regarding *B. velezensis* EM-A8 survival, we found that the population remains in the order of four logarithms after 30 days of application under natural field conditions. This shows that bacteria can colonize and remain viable in the phyllosphere of maize plants, which is your native environment, without modifying the microbial populations of the maize phyllosphere.

Keywords. biological control, maize, microbial communities, physiological profile

1. Introduction

Our research was carried out to select epiphytic microorganisms as potential biological control agents (BCA) against *Exserohilum turcicum, the* causal agent of northern leaf blight (NLB) in maize plants. Microbial isolates obtained from the maize phyllosphere were evaluated for their antagonistic activity (Sartori et al., 2015; 2017a). It is important to consider the relationship between biological interactions and environmental stress factors to achieve the selection of potential biological control agents (Nesci et al., 2005). In preliminary studies, we isolated possible antagonistic epiphytic microorganisms from leaves of maize with blight lesions. Most epiphytic populations consisted of bacteria, which were found in the order of 6 log CFU per gram of maize leaf (Sartori et al., 2015). *Bacillus velezensis* EM-A8 showed a dominance of the pathogen at a distance and a reduction of *E. turcicum* growth *in vitro* and *in plant* studies (Sartori et al., 2020).

Although BCA has been widely used to manage plant diseases, the ecological impact of their formulations on native phyllosphere bacteria and the relationship between phyllosphere microbial communities and plant health, remain poorly understood. The phyllosphere supports a tremendous diversity of microbes, which have the potential to influence plant biogeography and ecosystem function. Also, phyllosphere communities could be easily affected by exogenous factors, such as UV radiation, air pollution, and biological inoculation (Williams et al., 2013). Inoculation of BCA can alter the indigenous microorganisms within the phyllosphere, thereby affecting the community's ecological and functional properties (Zhang et al., 2008; Qin et al., 2019).

What happens in the phyllosphere with the application of BCAs in the phyllosphere is unknown in their interactions with the resident microbiome (Sylla et al., 2013). The success of biological control lies in the level of knowledge of the natural crop-pest-enemy system. Understanding the population dynamic balance between the organisms of the phyllosphere as an ecological system should lead to new approaches in agronomy.

The functional analysis of microbial communities involves the characterization of patterns and processes that allow predicting the stability of the ecosystem in the event of disturbances, clarifying the interactions between the organisms that compose it and the environment, and obtaining a diversity-function relationship. Their results constitute an important basis for understanding part of the relationships between carbon sources, nutrients, and formulations, with microbial activity in the ecosystem. Diverse studies of metabolizing profiles in the Biolog Ecoplate™ system, based on the similarity of the consumption of 31 carbon-sources, as a reflection of the catabolic potential of a community (Smalla et al., 1998; Garland, 1996), one approach to characterize the metabolic activity of bacteria is to analyze the pattern of carbon-source utilization. Ecoplates have been used to study the effects of different compounds and changes in physiological structure to microbial communities (Ros et al., 2006; Tiquia, 2010). To advance prevention strategies compatible with sustainable agriculture, it is necessary to deepen the knowledge of the foliar ecosystem. The microbiome of the leaf affects the natural aptitude of the populations of the plants and the quality of their productivity. In the phyllosphere, microorganisms compete for space and nutrient resources, both within each population and between populations. It is important to understand the composition of the phyllosphere microbiome, in particular the physiological properties of phyllosphere microorganisms and their role with respect to modulating population sizes of pathogens (Delmotte et al., 2009).

The aims of this study were: i) to detect changes that occur in the microbial community in maize phyllosphere after foliar application of *B. velezensis* EM-A8, and ii) to study the survival of antagonistic bacteria in the phyllosphere of maize leaves in the field. This will improve our understanding of the ecological function of the phyllosphere community on plants.

2. Effect of Formulation Application on the Microbial Community of Phyllosphere

Community-level physiological profiles with treatments carried out on three different conditions were determined: *In vitro*, field trial, and culture chamber. Treatments: *In vitro* 1) *B. velezensis* EM-A8. 2) *E. turcicum.* Field trial: 3) Control. 4) *B. velezensis* EM-A8 formulation (Sartori et al., 2025). 5) *B. velezensis* EM-A8 in NB. Culture chamber: 6) Control. 7) *B. velezensis* EM-A8 in NB. 8) *E. turcicum. 9*) *B. velezensis* EM-A8 in NB + *E. turcicum.*

2.1. *In Vitro*

Cultures of the antagonist *B. velezensis* EM-A8 plated on NA and the pathogen *E. turcicum* plated on glucose potato agar and incubated at 25°C.

2.2. Field Trial

Commercial maize seeds cultivar DK 7210 MG RR2 (Monsanto, Argentina) was used. This cultivar is resistant to insects and to the herbicide glyphosate. The field trials were carried out at a field in Río Cuarto, Córdoba (33°2'S latitude, 64°14'O longitude, 562 m altitude). Treatments were planted in three randomized blocks with three replications for each treatment. Individual plots consisted of two rows with 50 plants. Sowing took place during mid-January, while harvest was performed during mid-May. One application of the herbicide glyphosate was performed on the entire parcel 1 week after sowing according to common agricultural practices used for RR cultivars.

Plants at the V10 phenological stage (Ritchie and Hanway, 1982) were inoculated by foliar spraying with a formulation containing molasses and yeast extract (Sartori et al., 2025) and broth inoculum both with 10^7 CFU ml^{-1} *B. velezensis* EM-A8 to a_W 0,97 (Dallyl and Fox, 1980) using an atomizer. Samples corresponding to different treatments will be taken in the vegetative state R2, 28 days after the inoculation by spraying of the antagonist bacteria. In all treatments, leaves were cut with the presence of similar lesions of blight diseases; therefore, the 3 treatments consisted of a group of leaves with the presence of the pathogen. 300 mg of leaf sample will be placed in 30 ml of PBS (pH 7), they will be sonicated for 10 minutes (10 W) and then in a freezer at -20°C (Correa et al., 2007).

2.3. Culture Chamber

Commercial maize seeds cultivar DK 7210 MG RR2 (Monsanto, Argentina) was used. 3 seeds were sown per pot and 4 pots per treatment. *B. velezensis* EM-A8 (10^8 CFU/ml) inoculum (stage V4) was sprayed on 4-leaf seedlings. At seven days, *E. turcicum* (10^4 conidia/ml) was inoculated in the same way (Treatment 3 and 4). After seven days, a group of leaves from each treatment was made, they were placed in 300 mg in 30 ml of PBS in a tube, they were placed in a sonicator for 10 minutes (10 W), and then in a freezer at -20°C (Correa et al., 2007).

For the processing of the field and culture chamber samples, they were thawed and centrifuged for 10 minutes at 500 g at 4°C. Then, 150 μl of each supernatant sample was seeded in one well of the Ecoplate. Data collection used a Take3TM Epoch (Biotek Inc., Winooski, VT) reader through Gen5TM Microplate Data Analysis software (Biotek Inc.).

In Figure 1, the heat map obtained allows us to establish a relationship between the phyllosphere community and the metabolic profile. T5, T4, and T3, belonging to the field trial, were grouped in the dendrogram and showed the greatest capacity to use carbon sources. D-galacturonic Acid, Glucose-1-phosphate, Tween 80, Glycogen, and L-phenylalanine were the most used carbon sources by the microbial community. Samples of control plants (T3) and of plants treated with the formulation developed (T4) were grouped in the dendrogram. It is highlighted that the presence of the antagonist *B. velezensis* EM-A8 did not alter the metabolic pattern.

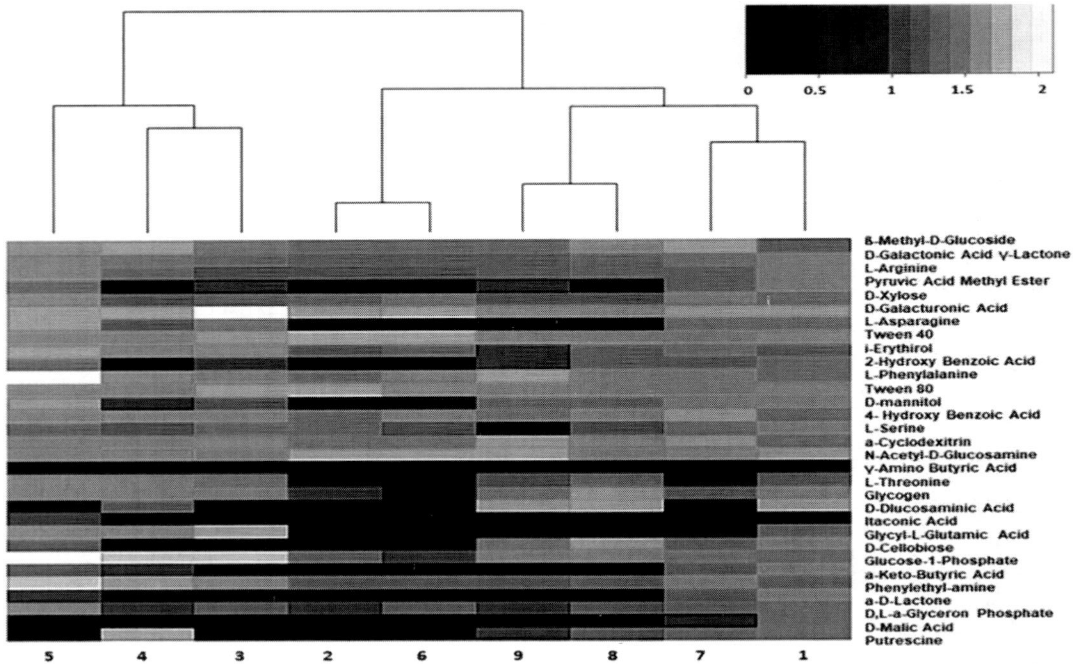

Figure 1. Comparative heat map of different treatments in Biolog ecoplates with the use of the 31 carbonated sources at 160 h of incubation. Treatments: *In vitro* 1) *B. velezensis* EM-A8. 2) *E. turcicum.* Field trial: 3) Control. 4) *B. velezensis* EM-A8 formulation. 5) *B. velezensis* EM-A8 in nutrient broth. Culture chamber: 6) Control. 7) *B. velezensis* EM-A8 in nutrient broth. 8) *E. turcicum*. 9) *B. velezensis* EM-A8 in nutrient broth + *E. turcicum.*

The pathogen *E. turcicum* (T2) and treatment control of plants in the chamber (T6) showed the least capacity to use carbonate sources and were grouped in the dendrogram.

Finally, treatments belonging to the chamber test, T9, T8, and T7, and the control treatment of *B. velezensis* EM-A8 (T1), are grouped in the dendrogram showing a similar ecological pattern, Glucose-1-phosphate, Tween 80, L-phenylalanine, and N-acetyl-D-glucosamine were the most used carbon sources.

Also highlighted that T7 (inoculated seedlings with the antagonistic) is grouped with its *in vitro* control, T1.

3. Survival of the Antagonist *B. Velezensis* EM-A8 on the Phyllophere

The purpose of the trial was to monitor the population of *B. velezensis* EM-A8. Survival and multiplication on the phylloplane were determined using antibiotic resistance as a marker. On the one hand, spontaneous mutants to 5% streptomycin were obtained by adapting the methodology of Krishna Kishore et al. (2005).

On the other hand, a trial was conducted in a commercial field in the rural area of Río Cuarto, Córdoba Province, Argentina. The trial was conducted with a randomized block design with three treatments: T1) control, T2) *B. velezensis* EM-A8, and T3) *B. velezensis* EM-A8 + fungicide (Amistar xtra gold). The cultivar DK 670 was used. Plants in the V8 stage were inoculated by foliar spraying with *B. velezensis* EM-A8 resistant to 5% streptomycin. Leaves from unsprayed plants were used as controls (T1). To determine the population of antibiotic-resistant *B. velezensis* EM-A8, samples were taken at different times and stages (at the time of inoculation (V8), 15, and 30 days later (VT and R1, respectively). The collected leaf samples were placed in phosphate buffer. They were then incubated for 1 h at 180 rpm and 30ºC. Serial dilutions were made, and aliquots were plated on media supplemented with streptomycin. The count was made after 48 h of incubation at 30ºC.

Table 1 shows the antagonist and total count in the phyllosphere at different stages of cultivation under natural field conditions. On the inoculation day, treatment *B. velezensis* EM-A8 count was 5,8 x 10^5 CFU ml^{-1}. 30 days after post-application of the inoculum, in the R1 stage, antagonistic bacteria remained in order of four logarithms (1,8 x 10^4 CFU ml^{-1}). It was found that *B. velezensis* EM-A8 can colonize and remain viable in the phyllosphere of maize plants.

In treatment 2, the total count remained in five logarithms to R1 (2,2 x 10^5 CFU ml^{-1}). Unlike treatment 3 (with fungicide application), which showed a decrease in the count in R1, both of *B. velezensis* (3 x 10^3 CFU ml^{-1}) and total count (7 x 10^4 CFU ml^{-1}).

The trial objective was to monitor the population of *B. velezensis* EM-A8 and to correlate it with foliar diseases. However, during the agricultural campaign and given the climatic conditions, a significant incidence of common rust (caused by *Puccinia sorghi*) was detected, but not of NCLB caused by *E. turcicum* at the time evaluated. However, NCLB was detected with a significant degree of severity in very late cultivars.

Table 1. Antagonist and total count in the phyllosphere at different stages of cultivation

Stage	Treatment	B. velezensis EM-A8 CFU ml^{-1} (NA + Streptomycin 5%)	Total count CFU ml^{-1} (NA)
V8	1	-	1,2 x 10^5
	2	5,8 x 10^5	9 x 10^5
	3	2,8 x 10^5	4,5 x 10^5
VT	1	-	2,1 x 10^5
	2	4,9 x 10^4	8 x 10^6
	3	9 x 10^3	3 x 10^5
R1	1	-	1,7 x 10^5
	2	1,8 x 10^4	2,2 x 10^5
	3	3 x 10^3	7 x 10^4

Treatments: 1) control. 2) *B. velezensis* EM-A8 (Streptomycin resistant). 3) *B. velezensis* EM-A8 (Streptomycin resistant) + fungicide (Amistar xtra gold).

3.1. Statistical Analysis

The analysis of variance (ANOVA) was made for growth parameters and viable counts using the program InfoStat version 2012 (Di Rienzo et al., 2012). Means were compared according to the DGC test ($p > 0.05$). The results obtained from the absorbance measured in the Biolog Ecoplates were analyzed through a non-linear regression (Abs. Alf * Exp), obtaining a dendrogram and heat map at 160 h of incubation, using Software OriginPro8 (OriginLab Corporation, Northampton, MA).

Discussion

Interactions between the host plant and the microbiota are based on the transfer of molecular and genetic information, and in that sense, it is known that some important colonization factors, such as secondary metabolites, QS systems, biofilm formation and cell signaling, are responsible for allowing this exchange of information (Braga et al., 2016; Flores-Núñez et al., 2020) therefore, all microorganisms capable of colonizing the phyllosphere respond to abiotic and biotic fluctuations involving different microbial mechanisms, such as the ability to extract nutrients, produce hormones and surfactants, motility and biofilm formation, can be key to the success of colonization (Nadakuduti et al., 2012; Ueda et al., 2018; Oso et al., 2019; Streletskii et al., 2019).

B. velezensis EM-A8 is an antagonistic bacterium native to the phyllosphere (Sartori et al. 2015; 2017a, b; 2020), and in our previous studies, antagonistic strains of the genus *Bacillus* spp., including *B. velezensis* EM-A8, demonstrated the ability to move and form biofilms under varying conditions of temperature, water potential, and nutrient availability (Fessia et al., 2022). Biofilms are aggregates of microbial communities in which cells adhere to each other and to a surface, enveloped in a matrix of extracellular polymeric compounds, which protects the community under adverse conditions (Davey and O'Toole, 2000). Bacterial cell aggregates must reach a minimum size for protection in unfavorable environments (Chaudhry et al., 2021), it is known that in nature, approximately 70% of the bacteria on leaves are found in aggregates, which gives them a selective advantage of survival and colonization over solitary cells on the leaf surface (Morris and Kinkel, 2002; Monier and Lindow, 2003), *B. velezensis* EM-A8 has that advantage.

The leaf surface is known to be a nutrient-limited environment, and its content is an important regulator of phyllosphere microbial communities (Yang et al., 2001). Phyllosphere bacteria have been reported to be primarily carbon-limited (Mercier and Lindow, 2000) and secondarily nitrogen-limited (Yadav et al., 2005). Temporal patterns of diversity can be sensitive indicators of ecosystem functioning and, at the same time, useful tools for its assessment. Therefore, functional analysis, as an indicator of the diversity and structure of microbial communities, allows determining the impact of the presence of BCA using different carbon sources, establishing a metabolic profile of the microorganisms and their behavior (Grayston and Campbell, 1996; Frąc et al., 2012). Our results when determining the metabolic profile of the phyllosphere microbial community showed a dendrogram in which the treatments belonging to the field trial were grouped. Plants with a naturally occurring presence of microorganisms demonstrated a similar ecological role for the microbial community present, exhibiting a greater capacity to utilize carbonate sources and indicating greater microbial diversity. Some research indicates that inoculation with BCA can alter the native microorganisms in the phyllosphere, thereby affecting the ecological and functional properties of the community (Zhang et al., 2008). Therefore, it is important to highlight that inoculation with the native bacterium *B. velezensis* EM-A8 at a high count (10^7 CFU/ml) did not alter the characteristic metabolic pattern of the maize plant phyllosphere.

Our results also showed that plants grown in a non-inoculated growth chamber (control) and the pathogen *E. turcicum* had the lowest capacity to utilize carbon sources. On the one hand, it is possible to interpret low microbial diversity for non-inoculated plants grown in a controlled atmosphere, such as a growth chamber, and on the other hand, a limited capacity for the pathogen to utilize carbonate sources over 160 h. Understanding this allows us to infer spatial and temporal variations in community structure, thus improving the sustainable management strategies we aim to develop against emerging diseases in the maize agricultural ecosystem.

Association between communities starts with initial adhesion to the leaf surface and ends with a complex network of interactions (Chaudhry et al., 2021). By spatially and temporally comparing the functional spectra of microbial communities, in this case a metabolic profile, it is possible to determine the differences between communities under the local or temporal effect of a BCA. This allows us to estimate the magnitude of the transformation induced by the agent and its consequences on the stability and sustainability of a particular ecosystem (Griffiths et al., 1997; Frąc et al., 2012). The data obtained demonstrate that the native foliar microflora did not show changes in microbial functional diversity following foliar inoculation of *B. velezensis* EM-A8, evaluated with different carbon sources. This is consistent with numerous studies indicating that the phyllosphere microbiota under field conditions was not significantly affected by introduced BCAs (Perazzolli et al., 2014; Wei et al., 2016) and contrasts with other research that showed that inoculation of *A. brasilense* can affect the response of the native bacterial community on tomato leaves (Correa et al., 2007).

Regarding the count of *B. velezensis* EM-A8 and total microorganisms in the phyllosphere, at different cultivation stages under natural field conditions, it was shown that 30 days after the application of the inoculum, the antagonistic bacteria remained in the order of 4 log, indicating the ability of the antagonistic bacteria to colonize and remain viable in the phyllosphere of maize plants, their native environment. This data is important, given that the stages evaluated include the critical period of the foliar disease that we intend to control, which is located 15 days before and 15 days after the flowering stage. Previously, the same monitoring of the selected bacteria in the phyllosphere had been carried out in a greenhouse trial, where it was determined that 39 days post-inoculation, *B. velezensis* EM-A8 remained at a population of 10^6 CFU/ml, the highest and most stable population of the selected bacteria (Sartori et al., 2017a). In the same sense, some researchers indicate that highly diverse leaf communities are negatively correlated with pathogen invasion and colonization (Purahong et al., 2018; Qin et al., 2019).

It is important to highlight that *B. velezensis* EM-A8 inoculation did not affect the total count of microorganisms in the phyllosphere, remaining at the same count as the control treatment. On the contrary, the total and antagonist counts were affected, decreasing by 1 log, when chemical fungicide was applied. In this sense, various investigations on the microbiome currently attempt to understand how microbial consortia in nature protect plants during disturbance, and the effects of immune priming (IP) through abiotic (applied chemical compounds) and biotic (biocontrol agents) stimuli seem to play an important role in managing abiotic stress tolerance and disease resistance (Kumar and Verma, 2018; Chaudhry et al., 2021). We agree on the need for further studies in this field of research, to explore the effects and underlying mechanisms of abiotic and biotic stress, and how they are transmitted to foliar microbial communities (Chaudhry et al., 2021).

Biological control represents an alternative to pesticide use and can also be implemented as a tool in integrated pest management in agroecosystems by leveraging inducible defenses (Grossi Vanacore et al., 2023). In this sense, it is possible to assert that the antagonistic bacterium *B. velezensis* EM-A8 can potentially be used to control maize diseases caused by *E. turcicum* and improve crop yield and production as part of a new biological management and breeding strategy.

Acknowledgments

Our thanks to the team of the microbial ecology laboratory of the Department of Ecological Sciences, Faculty of Sciences, University of Chile (Ñuñoa, Santiago de Chile) for their collaboration.

References

Braga, R. M., Dourado, M. N., Araújo, W. L. 2016. Microbial interactions: ecology in a molecular perspective. *Brazilian Journal of Microbiology*, 47(1), 86–98.

Chaudhry, V., Runge, P., Sengupta, P., Doehlemann, G., Parker, J., Kemen, E. 2021. Shaping the leaf microbiota: plant–microbe–microbe interactions. *Journal of Experimental Botany*, 72 (1) 36–56.

Correa, O. S., Romero, A. M., Montecchia, M. S., Soria, M. A. 2007. Tomato genotype and *Azospirillum* inoculation modulate the changes in bacterial communities associated with roots and leaves. *Journal of Applied Microbiology,* 102, 781–786.

Dallyl, H., Fox, A. 1980. Spoilage of material of reduced water activity by xerophilic fungi. *Technical Series* 15. London: Academic Press. p. 219–239.

Davey, M. E., O'toole, G. A. 2000. Microbial biofilms: from ecology to molecular genetics. *Microbiology and Molecular Biology Reviews,* 64, 847–867.

Delmotte, N., Knief, C., Chaffron, S., Innerebner, G., Roschitzki, B., Schlapbach, R., von Mering, C., Vorholt, J. A. 2009. Community proteogenomics reveals insights into the physiology of phyllosphere bacteria. *Proceedings of the National Academy of Sciences of the USA.* 106:16428–16433.

Di Rienzo, J., Casanoves, F., Balzarini, M., Gonzalez, L., Tablada, M., Robledo, C. 2012. InfoStat versión Argentina: GrupolnfoStat, FCA, Universidad Nacional de Córdoba. http://www.infostat.com.ar.

Frąc, M., K. Oszust y J. Lipiec. 2012. Community Level Physiological Profiles (CLPP), Characterization and Microbial Activity of Soil Amended with Dairy Sewage Sludge. *Sensors,* 12: 3253-3268.

Fessia, A., Sartori, M., García, D., Fernández, L., Ponzio, R., Barros, G., Nesci, A. 2022. *In vitro* studies of biofilm-forming *Bacillus* strains, biocontrol agents isolated from the maize phyllosphere. *Biofilm,* 4: 100097.

Flores-Núñez, V. M., Fonseca-García, C., Desgarennes, D., Eloe Fadrosh, E., Woyke, T., Partida-Martínez, L. P. 2020. Functional signatures of the epiphytic prokaryotic microbiome of agaves and cacti. *Frontiers in Microbiology,* 10, 3044.

Garland, J. L., 1996. Analytical approaches to the characterization of samples of microbial communities using patterns of potential C source utilization. *Soil Biology and Biochemistry,* 28 (2), 213–221.

Grayston, S., C. Campbell. 1996. Functional biodiversity of microbial communities in the rhizospheres of hybrid larch (*Larix europletis*) and Sitka Spruce (*Picea sitchensis*). *Tree Physiology,* 16:1031–1038.

Griffiths, B., Ritz, K., Wheatley, R. 1997. Relationship between Functional Diversity and Genetic Diversity in Complex Microbial Communities. In: *Microbial Communities. Functional vs. Structural Approach* (H. Insam y A. Rangger, Eds.). USA, Springer. 1:1-9.

Grossi Vanacore, M. F., Sartori, M., Giordanino, F., Barros, G., Nesci, A., García, D. 2023. Physiological Effects of Microbial Biocontrol Agents in the Maize Phyllosphere. *Plants*, 12, 4082.

Krishna Kishore, G., Pande, S., Podile, A. 2005. Biological control of late leaf spot of peanut (*Arachis hypogaea*) with chitinolytic bacteria. *Biological Control*, 95(10), 1157-1160.

Kumar, A., Verma, J. P. 2018. Does plant–microbe interaction confer stress tolerance in plants: a review? *Microbiological Research,* 207, 41–52.

Mercier, J., Lindow, S. E. 2000. Role of leaf surface sugars in colonization of plants by bacterial epiphytes. *Applied and Environmental Microbiology,* 66: 369–374.

Monier, J. M., Lindow, S. E. 2004. Frequency, size, and localization of bacterial aggregates on bean leaf surfaces. *Applied and Environmental Microbiology,* 70, 346–355.

Morris, C. E., Kinkel, L. L. 2002. Fifty years of phyllosphere microbiology: significant contributions to research in related fields. In: Lindow, S. E., Hecht-Poinar, E. I., Elliott, V., eds. Phyllosphere microbiology. St. Paul, MN: APS Press, 365–375.

Nadakuduti, S. S., Pollard, M., Kosma, D. K., Allen, C., Ohlrogge, J. B., Barry, C. S. 2012. Pleiotropic phenotypes of the sticky peel mutant provide new insight into the role of CUTIN DEFICIENT2 in epidermal cell function in tomato. *Plant Physiology,* 159, 945–960.

Nesci, A., Bluma, R., Etcheverry, M. 2005. *In vitro* selection of maize rhizobacteria to study potential biological control of *Aspergillus* section *Flavi* and aflatoxin production. *European Journal of Plant Pathology,* 113:159-71.

Oso, S., Walters, M., Schlechter, R. O., Remus-Emsermann, M. N. P. 2019. Utilisation of hydrocarbons and production of surfactants by bacteria isolated from plant leaf surfaces. *FEMS Microbiology Letters,* 366, fnz061.

Perazzolli, M., Antonielli, L., Storari, M., Puopolo, G., Pancher, M., Giovannini, O., Pindo, M., Pertot, I. 2014. Resilience of the natural phyllosphere microbiota of the grapevine to chemical and biological pesticides. *Applied Environmental and Microbiology,* 80(12):3585–3596.

Purahong, W., Orrù, L., Donati, I., Perpetuini, G., Cellini, A., Lamontanara, A., Michelotti, V., Tacconi, G., Spinelli, F. 2018. Plant microbiome and its link to plant health: host species, organs and Pseudomonas syringae pv. Actinidiae infection shaping bacterial phyllosphere communities of Kiwifruit plants. *Frontiers in Plant Science,* 9, 1563.

Qin, C., Tao, J., Liu, T., Liu, Y., Xiao, N., Li, T., Gu, Y., Yin, H., Meng, D. 2019. Responses of phyllosphere microbiota and plant health to application of two different biocontrol agents. *AMB Express*, *9*(1), 42.

Ritchie, S., Hanway, J. 1982. *How a corn plant develops.* Iowa State University of Science and Technology. Ames, Iowa. Special Report N°48.

Ros, M., Goberna, M., Moreno, J. L., Hernandez, T., Garcia, C., Insam, H., Pascual, J. A. 2006. Molecular and physiological bacterial diversity of a semi-arid soil contaminated with different levels of formulated atrazine. *Applied Soil Ecology,* 34 (2–3), 93–102.

Sartori, M., García, D., Grossi Vanacore, M. F., Fessia, A., Nesci, A. 2025. Biofungicide formulation based on *Bacillus velezensis* EM-A8 for control of maize foliar diseases. *Rev. FCA UNCuyo*. ISSN (online) 1853-8665. In press.

Sartori, M., Bonacci, M., Barra, P., Fessia, A., Etcheverry, M., Nesci, A. and Barros, G. 2020. Studies on possible modes of action and tolerance to environmental stress conditions of different biocontrol agents of foliar diseases in maize. *Agricultural Sciences*, 11, 552-566.

Sartori, M., Nesci, A., Garcia, J., Passone, M., Montemarani, A., Etcheverry, M. 2017 (a). Efficacy of epiphytic bacteria to prevent northern leaf blight caused by *Exserohilum turcicum* in maize. *Revista Argentina de Microbiología,* 49(1):75-82.

Sartori, M., Nesci, A., Montemarani, A., Barros, G., Garcia, J, Etcheverry, M. 2017 (b). Preliminary evaluation of biocontrol agents against maize pathogens *Exserohilum turcicum* and *Puccinia sorghi* in field assays. *Agricultural Sciences*, 8: 1003-1013.

Sartori, M., Nesci, A., Formento, A., Etcheverry, M. 2015. Selection of potential biological control of *Exserohilum turcicum* with epiphytic microorganisms from maize. *Revista Argentina de Microbiología,* 47(1):62-71.

Smalla, K., Wachtendorf, U., Heuer, H., Liu, W., Forney, L. 1998. Analysis of BIOLOG GN substrate utilization patterns by microbial communities. *Applied Environmental and Microbiology*, 64, 1220–1225.

Streletskii, R. A., Kachalkin, A. V., Glushakova, A. M., Yurkov, A. M., Demin, V. V. 2019. Yeasts producing zeatin. *PeerJ* 7, e6474.

Sylla J., Alsanius B., Krüger E., Reineke A., Bischoff-Schaefer M., Wohanka W. 2013. Introduction of *Aureobasidium pullulans* to the phyllosphere of organically grown strawberries with focus on its establishment and interactions with the resident microbiome. *Agronomy* 3, 704-731.

Tiquia, S. M., 2010. Metabolic diversity of the heterotrophic microorganisms and potential link to pollution of the Rouge River. *Environmental Pollution,* 158 (5), 1435–1443.

Ueda, H., Kurose, D., Kugimiya, S., Mitsuhara, I., Yoshida, S., Tabata, J., Suzuki, K, Kitamoto, H. 2018. Disease severity enhancement by an esterase from non-phytopathogenic yeast *Pseudozyma Antarctica* and its potential as an adjuvant for biocontrol agents. *Scientific Reports,* 8, 16455.

Wei, F., Hu, X., Xu, X. 2016. Dispersal of *Bacillus subtilis* and its effect on strawberry phyllosphere microbiota under open field and protection conditions. *Sci Rep.* 6, 22611.

Williams, T., Moyne, A., Harris, L., Marco, M. 2013. Season, irrigation, leaf age, and *Escherichia coli* inoculation influence the bacterial diversity in the lettuce phyllosphere. *PLoS ONE* 8(7): e68642.

Yadav, R.; Karamanoli, K.; Vokou, D. 2005. Bacterial Colonization of the Phyllosphere of Mediterranean Perennial Species as Influenced by Leaf Structural and Chemical Features. *Microbial Ecology*, 50, 185–196.

Yang C., Crowley, D., Borneman, J., Keen, N. 2001. Microbial phyllosphere populations are more complex than previously realized. *Proceedings of the National Academy of Sciences*, 98 (7) 3889-3894.

Zhang B., Bai Z., Hoefel D., Tang L., Yang Z., Zhuang G., Yang J., Zhang H. 2008. Assessing the impact of the biological control agent *Bacillus thuringiensis* on the indigenous microbial community within the pepper plant phyllosphere. *FEMS Microbiol Lett,* 284(1):102–108.

Chapter 3

Physiological and Genetic Aspects Influenced by Light Quality in Two Biocontrol Agents, *Bacillus subtilis* EM-A7 and *Bacillus velezensis* EM-A8

Aluminé Fessia*
Germán Barros
and Andrea Nesci†
Laboratorio de Ecología Microbiana Ambiental. Instituto ICBIA, (CONICET-UNRC), Departamento de Microbiología e Inmunología, Facultad de Ciencias Exactas, Físico-Químicas y Naturales, Universidad Nacional de Río Cuarto, Río Cuarto, Córdoba, Argentina

Abstract

Maize (Zea mays L.), a major crop in Argentina and a staple food around the world, is affected by the emergence and re-emergence of foliar diseases. The main pathogens found on maize leaves are Puccinia sorghi and Exserohilum turcicum, responsible for common rust and northern corn leaf blight (NCLB), respectively. Different crop protection strategies have come up as part of transformational agriculture in a bid to prevent and/or control the appearance of pests in a given ecosystem. Such strategies include crop rotation and the use of resistant cultivars; however, chemical pesticides are the most widely used approach against many plant pathogens. An emerging alternative is using living organisms for human good, i.e., microbial biocontrol agents, to combat pests,

---

* Corresponding Author's Email: afessia@exa.unrc.edu.ar
† Corresponding Author's Email: anesci@exa.unrc.edu.ar

In: Biocontrol Agents (BCAs)
Editor: Daiana García
ISBN: 979-8-90134-102-5

pathogens, and plant pathogen populations. Microorganisms isolated from the phyllosphere have been found to help decrease the number of foliar pathogens on the surface of leaves, i.e., pathogens inhabiting the same niche, which is ecologically significant. Bacteria tend to live in biofilms, which are communities of aggregated cells embedded in a self-produced extracellular polymeric matrix (ECM). In nature, approximately 70% of bacteria exist as aggregates on leaf surfaces, which confers them a selective advantage for survival and colonization. In our laboratory, previously, Bacillus subtilis EM-A7 and Bacillus velezensis EM-A8 were selected as potential biocontrol agents against E. turcicum, and their modes of action and tolerance to different environmental conditions were analyzed. More recently, we have focused on evaluating the capacity to form biofilm and motility of both strains under different environmental conditions using diverse microbiological techniques. We detected that both are capable of motility and biofilm formation under changing conditions regarding temperature, water potential, and nutrient availability. In addition, our results show that growth rate, the formation of complex colonies and pellicles, biofilm quantification, motility, and antagonism towards E. turcicum evaluated in EM-A7 and EM-A8 were influenced by light quality, i.e., wavelength and specific photosynthetically active photon flux density (PPFD), as well as by water potential and nutrient availability. These results should allow us to have a more accurate picture of the ecology of microorganisms in the phyllosphere and of their potential uses in agricultural practices.

Keywords: maize, biocontrol agent, biofilm, motility, phyllosphere

1. Introduction

Over the past fifty years, the increase in food demand brought about by the exponential growth of the global population signified a shift in cropping systems, from farms with high genotypic diversity to the predominance of genetic monocultures. As the population continues to expand and climate change intensifies, agriculture will face substantial challenges in sustainably increasing productivity to feed an estimated 9 billion people by 2050 (Lahlali et al., 2022; Röös et al., 2017). One of the major obstacles to achieving food security is plant diseases, which significantly reduce crop yields and cause serious economic losses (Adivitiya & Khasa, 2017). Although chemical pesticides are the most widely used approach against most pathogens causing these diseases (Adivitiya & Khasa, 2017; Carmona & Sautua, 2016; Formento,

2010), several prevention and control alternatives have come up as part of transformational agriculture, including crop rotation (Hill, 1998; Karp et al., 2015) and the use of cultivars that are resistant to major diseases (Tandzi & Mutengwa, 2020).

Maize (*Zea mays* L.) is a staple food in most human diets, as well as crucial for animal feed and biofuel production. It is grown in 13% of cropped areas around the world and accounts for 39% of all cereal production, which makes its annual production the highest for any cereal crop (Berger, 2024). An estimated one billion tons were produced globally in 2021-2022 (FAOSTAT, 2024), which means a 3.2% yearly increase since 1972 (Knoema, 2023). This is more than the 2.4% increase required per year to meet the expected demand by 2050 (Ray et al., 2013). In Argentina, approximately 46 million tons were cultivated on an area of 7.8 million hectares in 2024-2025 (BCR, 2025).

Biotic factors are responsible for 23% of maize yield losses every year (Savary et al., 2019). Emerging and re-emerging foliar diseases, in particular, greatly affect the crop at all stages of development (Formento, 2010). This is compounded by the increased adoption of conservation (i.e., no tillage) agriculture (Jat et al., 2021), which creates favorable conditions for crop pathogens that persist in the soil, including some foliar pathogens, to cause severe global outbreaks. Some of the most common foliar diseases in maize are northern corn leaf blight (NCLB), gray leaf spot, and rust (Berger, 2024), all caused by fungi. Argentina is no stranger to the global trend: climate change and alterations in production models have led to the re-emergence of diseases affecting maize, especially foliar ones (De Rossi, 2017). Two are the most frequent at a national level: common rust and NCLB. The former, caused by *Puccinia sorghi*, is endemic and occurs in all seasons, with varying degrees of intensity according to hybrid susceptibility, pathogen biotypes, and environmental conditions during the crop cycle (Darino et al., 2018). As for NCLB, its causative agent is the asexual form of hemibiotrophic Dothideomycete *Exserohilum turcicum* (Pass.) K.J. Leonard & Suggs (Leonard, 1974). The visible signs of the disease are greyish lesions that start as chlorotic flecks and mature into elliptical or cigar-shaped lesions, 2.5 to 30 cm in length (White, 1999). The fungus invades hosts through the epidermis, blocks vascular tissues (Kotze et al., 2019), causes plant lodging, and reduces the photosynthetic leaf area. In cases of severe infection during silking and grain filling, it may be responsible for between 30 and 90% of yield losses (Berger et al., 2020; Jindal et al., 2019).

Management of NCLB and other fungal diseases in maize usually relies on chemical formulations based on triazoles, which are demethylation inhibitor fungicides, and strobilurins, which are quinone outside inhibitors (FRAC, 2016). For instance, products available in the market combine trifloxystrobin+tebuconazole (Nativo; Bayer), pyraclostrobin+epoxiconazole (Opera; BASF), and azoxystrobin+cyproconazole (AMISTAR XTRA; Syngenta). Even though fungicides can reduce the severity and epidemic rate of disease and thus ensure good yields (Carmona & Sautua, 2016; Couretot et al., 2012), the repeated and exclusive use of at-risk products (Smith & Proost, 2020) makes it more likely for the frequency of resistant strains in a given pathogenic population to outnumber that of sensitive individuals. In other words, inappropriate application may lead to the development of microbial resistance and thus to disease re-emergence. In addition, it can pollute water and degrade the soil (Garg & Smith, 2014). Integrated management schemes attempt to bypass these issues by replacing or complementing pathogen suppression through chemical control with other strategies. Disease may be prevented by reducing the initial inoculum of causal pathogens in the field through tillage, rotation with non-host crops, and manipulation of environmental factors (Devi et al., 2020). The implementation of biological control is also gaining traction.

According to Stenberg et al. (2021), biological control is "...the exploitation of living agents (including viruses) to combat pestilential organisms (pests and pathogens), directly or indirectly, for human good." A non-native biocontrol agent (BCA) could become invasive and have a deleterious impact on the environment where it is introduced (Singh et al., 2020). In contrast, if the BCA has been isolated from the same ecosystem inhabited by the target pathogen, its reintroduction does not seem to pose significant risks (Alabouvette et al., 2012). In addition to being safe, it may have adaptive ecological advantages that make it effective in limiting adverse effects on plant health and productivity in that ecosystem (Hashem et al., 2019). In the case of BCAs aimed at antagonizing foliar pathogens, the ecosystem in question is the phyllosphere.

Morris (2002) defined the phyllosphere as "the microenvironment extending from the leaf surface outward to the outer edge of the boundary layer surrounding the leaf and inward into the leaf tissues," and Sivakumar et al. (2020) as "the aerial part of the plant or the parts of a plant above the ground, usually surface of leaves, considered a habitat for microorganisms." This ecological niche is inhabited by a wide variety of epi- and endophytic microorganisms (i.e., living on and within leaves). The total area occupied by the phyllosphere of plants on Earth is estimated to exceed 1 billion square kilometers and could harbor up to 10^{26} bacterial cells (Vorholt, 2012).

Microbes in the phyllosphere develop intimate associations with leaves, as well as adaptive traits which allow them to withstand changes in temperature, light, UV radiation, and water and nutrient availability (Bashir et al., 2022; Thapa & Prasanna, 2018). Their influence on host plants may be non-existent, pathogenic, or beneficial (Bashir et al., 2022). Non-pathogenic species can prevent pathogenic colonization, decrease stress, and positively impact plant growth and fitness (Sivakumar et al., 2020; Vorholt, 2012). Thus, it is from among this group that BCAs may be isolated and then inoculated onto leaves to help plants combat pathogens from the same environment (Bashir et al., 2022; Hashem et al., 2019). Their modes of action include outcompeting pathogens for nutrients and space, disrupting their communications, secreting metabolites or enzymes against them, or even parasitizing them.

Light wavelength is a significant determinant of the dynamics and functioning of leaf-associated microbial communities. (Alsanius et al., 2019, 2021; Gharaie et al., 2017). Studies have shown that many processes and aspects of non-phototrophic bacteria are influenced by light quality or the absence of light. These include substrate utilization for metabolic activity (Gharaie et al., 2017; Alsanius et al., 2021), respiration, biofilm formation (Karlsson et al., 2023), growth rate, motility, and the production of secondary metabolites, which are regulated by light (Oberpichler et al., 2008; Yu & Lee, 2013; Wu et al., 2013). Moreover, six distinct photosensory proteins have been identified in non-phototrophic bacteria, each with a range of absorption spectra spanning both the visible and non-visible light spectrum (Alsanius et al., 2024). These are phytochromes (PHY), photoactive yellow proteins (PYP), rhodopsins, blue light-absorbing proteins cryptochrome/photolyase (Cry/PHR), blue light-using flavin adenine dinucleotide (BLUF-FAD), and light oxygen voltage (LOV) domain (Gomelsky & Hoff, 2011; Wu et al., 2013).

Despite the abundant evidence of light's importance in the ecology and physiology of microorganisms in the phyllosphere, its effects on BCAs have not been sufficiently investigated. For BCAs to be successfully (re)introduced on leaves, they must spread on the surface, attach themselves to it, and exhibit antagonism (Alsanius et al., 2021). Since these steps are dependent on many of the processes mentioned above (motility, biofilm formation, production of metabolites), light is likely to have an impact on the effectiveness of a given BCA against a foliar pathogen. This chapter summarizes the results of our research into the effects of different light wavelengths on a number of physiological and genetic aspects of *Bacillus subtilis* and *Bacillus velezensis*, two potential BCAs for NCLB. Such aspects comprise growth rate, the formation of complex colonies and pellicles, biofilm quantification, motility, antagonism toward *E. turcicum*, and relative gene expression.

2. Selection of Biocontrol Agents

For several years, our laboratory has been conducting research into potential biocontrol agents that may prevent and/or mitigate the negative impact of phytopathogens on maize. *Bacillus subtilis* EM-A7 and *Bacillus velezensis* EM-A8 were isolated from maize leaves with blight lesions in fields planted with three cultivars in Chucul, Río Cuarto, and Vicuña Mackenna (Córdoba, Argentina). They showed a dominance of *E. turcicum* at a distance and significantly reduced its growth rate in vitro, by about 84 and 98 percent at water potential values (Ψ) of − 1.38 MPa and − 4.19 MPa (Sartori et al., 2015). In the greenhouse, they reduced NCLB severity at similar rates for the first 25 days after inoculation on maize plants, although the greatest effect was observed after day 29. The lowest incidence was recorded on day 20 and corresponded to the treatment with the strain EM-A8. High population sizes were registered for both strains in the phyllosphere for 39 days after inoculation (Sartori et al., 2017). Under field conditions, the reduction of NCLB severity was higher by over 50% for 40 days in plants at stages R2 to R4 treated with the *Bacillus* spp. (Sartori et al. 2017). The strains were subsequently observed to produce high levels of protease and β-1,3-glucanase, and to be tolerant to fluctuations in UV radiation, temperature, and osmotic stress. This suggests they are likely to have high survival rates on leaf surfaces (Sartori et al., 2020).

The bacteria were identified following Bergey's Manual of Systematic Bacteriology at the genus level, and through biochemical and molecular methods at the species level. These included an API 50 CH kit (bioMérieux, Lyon, France) and sequencing of 16S–23S rRNA (GenBank accession number: OL704804-OL704805), coupled with MALDI-TOF MS (Fessia et al., 2022; Sartori et al., 2015, 2020). The two draft genomes were recently sequenced and annotated. Several genes were found that encode antimicrobial activity and properties that may enhance survival on maize leaves (Fessia et al., 2024).

Furthermore, the biofilm-forming ability and motility of EM-A7 and EM-A8 were evaluated *in vitro* under varying conditions of temperature, water potential, growth medium, exposure to light, and incubation time (Fessia et al., 2022; Fessia et al., 2024). These changes affected the robustness of their biofilm, as demonstrated by the morphological phenotypes of their colonies and pellicles. Colonies formed by these two strains can have many morphological phenotypes. They may be creamy and convex and have lobed, irregular edges and numerous wrinkles and veins spreading from the center to the edges. In other cases, colonies appeared less complex-small, smooth, and round, with well-defined edges. Both strains exhibited motility, including swimming and swarming, under the conditions tested. Custom design of led sources

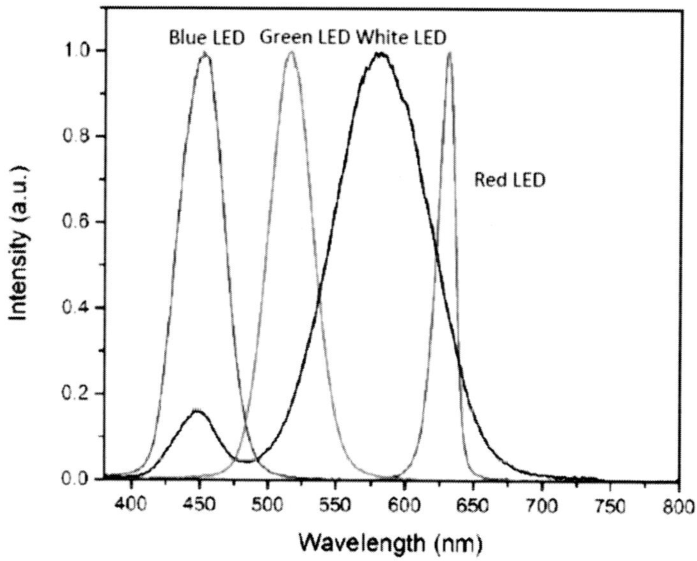

Figure 1. Normalized emission spectrum of each LED source used.

To evaluate the specific effects of light quality on certain physiological parameters of EM-A7 and EM-A8, LED devices were designed to simulate field conditions and thus approximate the results as closely as possible to what occurs in real-world scenarios. Each light source consisted of a panel of LEDs within a given wavelength spectrum and at a specific photosynthetically active photon flux density (PPFD). The panels were embedded in plastic boxes 32 cm in width, 43 cm in length, and 17 cm in height, with a 17 L capacity. Each box also contained a heat-dissipating plate and a cooling system. Four boxes were prepared, one for each of the following spectra: blue (417–486 nm, maximum emission at 452 nm), green (482–553 nm, maximum emission at 517 nm), red (wavelength range from 610 to 643 nm, with maximum emission at 631 nm), and white (432–658, maximum emission at 582 nm). The spectra were measured with a spectroradiometer (Vernier, SpectroVis Plus) (Figure 1).

Light intensity was maintained at 460 µmol m^{-1} s^{-2}, as indicated by reference radiation values. These were average measurements made by the Weather Station at the Facultad de Agronomía y Veterinaria (Universidad Nacional de Río Cuarto), between November and March of 2019 and 2022, in a field destined for academic research. The values, expressed as solar radiation (MJ m^{-2} d^{-1}), were converted into micromoles of photons per square meter per second (Reis & Ribeiro, 2020).

3. Results

3.1. The Effect of Light Quality on Biofilm Formed in Vitro

A series of qualitative and quantitative techniques was performed in our laboratory to study biofilm formation by EM-A7 and EM-A8 *in vitro*. One of them consisted of identifying differences in the architectural complexity of colonies and pellicles formed by the strains in the dark and under each LED source, after being incubated at room temperature on two media: maize leaves agar (MLA) and MSgg agar (a biofilm-inducing medium) (Fessia et al., 2022).

Figure 2 shows the biofilm development in the form of architecturally complex colonies by *Bacillus subtilis* EM-A7 and *Bacillus velezensis* EM-A8 under dark and green, red, white and blue light. In the dark (control), the colonies formed by the two strains grown on MSgg at Ψ=-1.38 MPa had numerous wrinkles and veins sprouting from the center toward the edges. The center was creamy and convex, and the edges lobed and irregular. On the same medium at Ψ= -4.09 MPa, they showed fewer wrinkles and veins. On MLA at Ψ=-1.38 MPa, the strains produced ring-shaped colonies from the edges to the center, with a smooth, concave center and lobed, irregular edges. On the same medium at Ψ=-4.09 MPa, EM-A7 colonies had a more defined center and fewer irregular edges than EM-A8 colonies. In general, control colonies were less complex when they developed at a lower water potential.

Exposure to green and red light led mostly to the formation of macrocolonies, which were smaller than the control but quite similarly characterized, regardless of the medium or water potential. On MLA, colonies were ring-shaped from the edges to the center; this center was smooth and concave; the edges were irregular but less complex than in the control. On MSgg at Ψ = -1.38, the colonies of both strains did not differ remarkably from the control. The exception was MSgg at Ψ = -4.09 MPa, on which EM-A7 grew very weakly, and EM-A8 did not grow at all.

Under white light, the colonies of both strains on MLA at Ψ=-1.38 MPa were ring-shaped from the edges to the center, much like the control. On MSgg at Ψ = -1.38 MPa, they were smaller and had fewer wrinkles and veins than the control, but their edges were clearly defined. Neither strain formed colonies in the presence of blue light, i.e., this quality of light inhibited biofilm development in the form of architecturally complex colonies in both media at both water potentials.

The Congo red dye in the strains grown in the control treatment, that is, in darkness in both culture media and water potentials, showed that these bacteria synthesized a significant number of exopolysaccharides, which coincided with biofilm formation in the form of complex colonies under the effect of the different LED lights. The intensity of the dye's color was observed to vary according to biofilm formation.

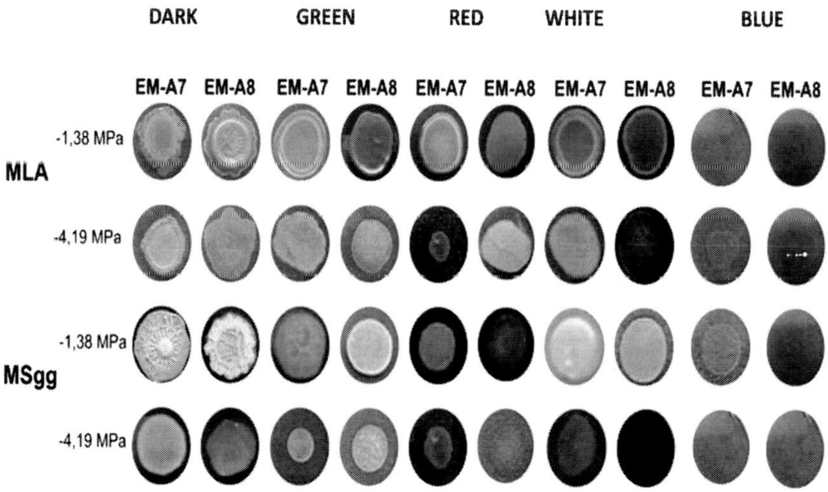

Figure 2. Biofilm development in the form of architecturally complex colonies by *Bacillus subtilis* EM-A7 and *Bacillus velezensis* EM-A8 under four light sources at 460 μmol m^{-1} s^{-2} after being incubated at room temperature on MLA and MSgg agar at different water potential values, supplemented with 40 μg/mL of Congo Red. MSgg: biofilm-inducing medium; MLA: maize leaves agar; water potential values Ψ=−1.38, −4.19 MPa.

Figure 3 shows the development of biofilm in the form of pellicles after the strains grew under exposure to the four LED sources in MSgg broth and maize leaves broth (MLB) at Ψ =-1.38 and -4.09 MPa.

In the case of EM-A7, the pellicles formed under green or red light in either medium at Ψ = -1.38 were similar to those in the control: densely wrinkled, robust, and thick. Under white light, pellicles were visible but weak and lacked distinctive features. There was no pellicle formation in MLB at the lower water potential. Similarly, EM-A7 did not produce pellicles under green or white light in MSgg at Ψ =-4.19 MPa. In the same medium, red light was linked to the production of fewer complex pellicles, without a distinctive macroscopic architecture. Finally, the strain did not form pellicles under blue light, regardless of the other experimental conditions.

As for EM-A8, pellicle complexity after exposure to green light in MLB and MSgg at Ψ = -1.38 MPa was similar to the control. Under red light in both media at Ψ = -1.38, pellicles were less complex and had no distinctive macroscopic structure. Moreover, the strain appeared more fragile to the naked eye. No pellicles were observed under red light in either medium at Ψ = -4.09 MPa, nor under white or blue light in general.

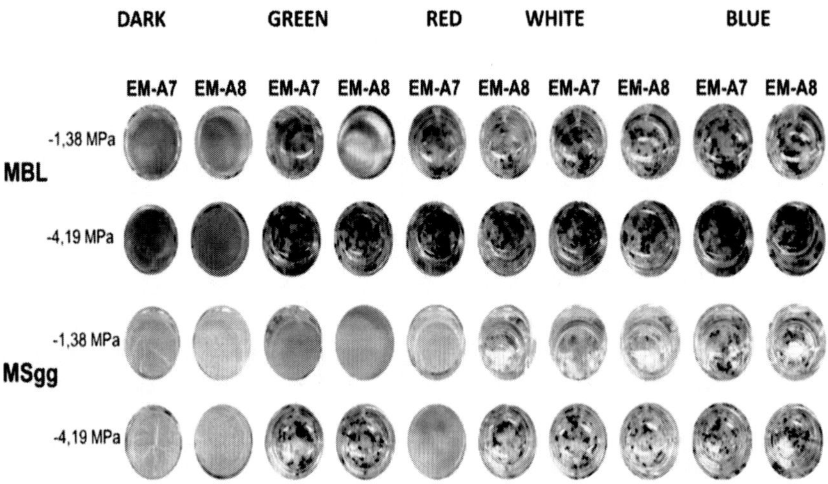

Figure 3. Biofilm development in the form of architecturally complex pellicles by *Bacillus subtilis* EM-A7 and *Bacillus velezensis* EM-A8 under four light sources at 460 μmol m^{-1} s^{-2}, after being incubated at room temperature on MLA and MSgg agar at different water potential values, MSgg: biofilm-inducing medium; MLB: maize leaves broth; water potential values Ψ=−1.38, −4.19 MPa.

The structure and spatial organization of the biofilm developed under the different conditions were explored further through scanning electron microscopy (SEM). The colonies were characterized by their three-dimensional structure and the presence of EPSs. The degree of architectural complexity observed on agar *in vitro* was consistent with SEM observations of structurally complex colonies, and the differences recorded after exposure to each light were likewise corroborated by SEM (Figure 4). In general, the morphology of EM-A7 and EM-A8 bacilli, and their distribution into clusters submerged in an amorphous substance, were similar for up to 48 h after exposure to the LED sources, no matter the specific light quality. After exposure to white, red, or green light on MLA and MSgg agar at Ψ =-1.38 MPa, the bacterial cells appear dispersed in a compact biofilm matrix, which is observable as rough material. Such rough material suggests the presence of EPSs on the surface and is particularly abundant in the images taken after exposure to white light on MLA and MSgg at either water potential.

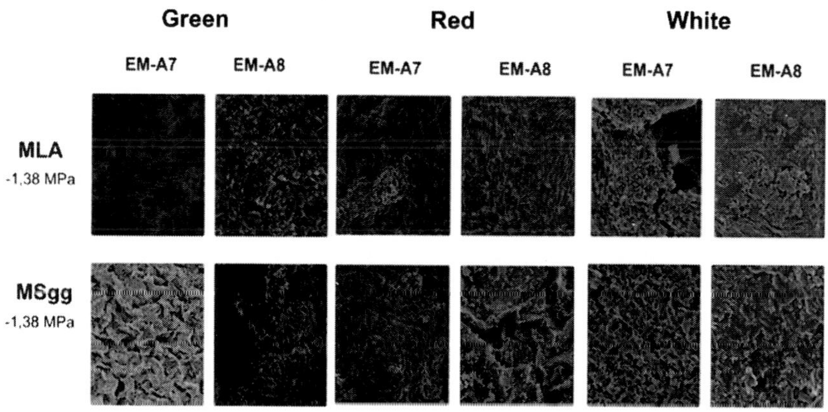

Figure 4. Scanning electron micrographs of colonies formed by *Bacillus subtilis* EM-A7 and *Bacillus velezensis* EM-A8 on MLA and biofilm-inducing MSgg agar at $\Psi = -1.38$ MPa, after 48 h of incubation at room temperature.

Table 1 shows the relative significance of four factors (growth medium, water potential, LED source, strain) and the interactions between them in the quantification of biofilm (OD560nm). An analysis of variance demonstrated that all the factors and their interactions were influential, but the most important was growth medium (GM) ($F = 500.51$, $p<0.0001$), followed by LED source ($F=219.94$, $p<0.0001$) and water potential (Ψ) ($F=179.59$, $p<0.0001$). The least influential was strain (S) ($F=0.84$, $p<0.0001$).

Table 1. Relative significance of growth medium, water potential, LED source, strain, and their interactions in the quantification of biofilm (OD560nm) formed by EM-A7 and EM-A8

Factor[a]	SC[b]	gl	CM	F	P
GM	38.19	1	38.19	510.51	<0.0001
Ψ	13.44	1	13.44	179.59	<0.0001
L	65.82	4	16.46	219.94	<0.0001
S	0.06	1	0.06	0.84	<0.0001
GM* Ψ	32.72	1	32.72	437.27	0.3597
GM*L	13.3	4	3.33	44.45	<0.0001
GM*S	0.25	1	0.25	3.29	<0.0001
Ψ*L	22.09	4	5.52	73.82	0.0717
Ψ*S	0.36	1	0.36	4.76	<0.0001
L*S	5.45	4	1.36	18.22	0.0305
GM* Ψ*L	11.01	4	2.75	36.8	<0.0001
GM* Ψ*S	0.06	1	0.06	0.77	<0.0001
GM* L*S	2.53	4	0.63	8.47	0.38
Ψ *L*S	2.33	4	0.58	7.79	<0.0001
GM* Ψ*L*S	2.76	4	0.69	9.24	<0.0001

[a] GM: growth medium; Ψ: water potential; L: LED source; S: strain. [b] SC: means of the squares; gl, degrees of freedom.

Biofilm formation by EM-A7 was significantly influenced by the LED source ($p<0.0001$), in both media at both water potential values. The highest production was found in MSgg broth at $\Psi=-1.38$ MPa under red, green, and white light. Here, the values were similar to those of the control. The lowest production was observed in MLB at $\Psi=-1.38$ MPa. Production under blue light was significantly lower than in the control or under other light qualities.

The LED source also had a significant impact on biofilm formation by EM-A8 ($p<0.0001$). The highest production occurred in the dark and was significantly higher than under any of the LED sources in MSgg broth and MLB at $\Psi=-1.38$ and MLB at $\Psi=-4.09$ MPa. In MSgg at $\Psi=-1.38$ MPa, the OD560nm values obtained under red and white light were similar to the control. In general, the exposure of either strain to red or white light resulted in biofilm formation values similar to those achieved in the dark in a biofilm-inducing medium. On the other hand, blue light overall inhibited biofilm formation.

3.1. The Effects of Light Quality on Motility

The effect of light quality on the motility of EM-A7 and EM-A8 was assessed on nutritive agar (NA) 0.3% for swimming and NA 0.7% for swarming. The strains were incubated for 48 h and subjected to a 13 h light (white, red, blue, or green)/11 h dark photoperiod regime.

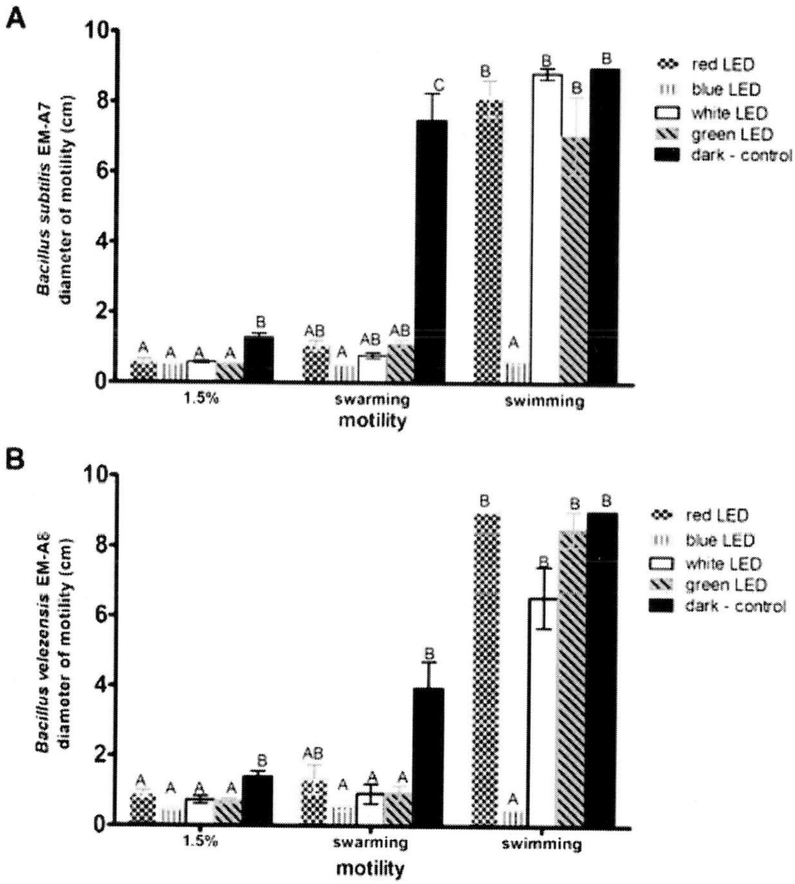

Figure 5. *Bacillus* strains' motility. (A) Means of the diameter of colonies formed by *Bacillus subtilis* EM-A7 on NA (nutritive agar) 1.5% (control), 0.7% (swarming) and 0.3% (swimming). (B) Means of the diameter of colonies formed by *Bacillus velezensis* EM-A8 on NA 1.5%, 0.7% and 0.3%, after 48 h of incubation at room temperature under different light sources. Mean of disc diameter ± SEM from four experiments performed in duplicate. The nominal p-value for statistical significance was $p<0.05$, indicated by different letters.

Motility was considered to take place when the diameter of the colony formed by a strain was at least 20% bigger than that of the control colony, formed on NB fortified with 1.5% agar (Bartolini & Grau 2019). As shown in Figure 5, A and B swimming by both EM-A7 and EM-A8 were significantly reduced ($p = 0.0355$ and $p = 0.0135$, respectively) after exposure to blue light, whereas the colonies formed under red, white, and green light were similar in diameter to the control. In other words, under these three light qualities, the strains were able to maintain their swimming ability. In addition, the colony diameters of each strain were 20% higher than those observed in the 1.5% agar medium, indicating bacterial motility. On the other hand, light had a much more remarkable effect on swarming by both strains ($p = 0.0062$ for EM-A7 and $p = 0.0228$ for EM-A8). This ability decreased significantly compared to the control, with no differences between the strains. In fact, the diameter of the colonies formed under the different light qualities was no greater than 20% of the diameter of control colonies.

Motility in the controls, for its part, differed significantly depending on the light source.

3.2. The Effect of Light Quality on Growth

The effect of light on the growth of the two strains was assessed in NB at room temperature for eight hours. Two parameters were examined: generation time (g) and constant growth rate (k). The results are shown in Figure 6.

The lag phase for EM-A7 was estimated at approximately three hours in the control (incubated in the dark) and after exposure to red, green and white light. The exponential phase was four hours long, and the stationary phase was never reached in the duration of the assay. Therefore, these three light qualities did not affect bacterial growth under the chosen experimental conditions. In contrast, exposure of EM-A7 to blue light led to an increase in optical density after five hours of incubation, which suggests that this light quality negatively affects growth. The values for *k* were similar to the control under red or green light, but significantly lower under white light. Those for *g were similar to the control* under red, green and white light.

The lag phase of EM-A8 showed more variation under the different LED sources and with respect to the control. It lasted three hours in the dark and under green light, and four hours under red and white light. As with the other strain, blue light negatively affected growth, as evidenced by the increased optical density values after five hours of incubation. The exponential phase was longer under red light than under other light sources or in the dark. In EM-A8, red light enhanced growth (*k*) compared with the control, while green and white light led to values similar to the control. By contrast, blue light significantly reduced k in both strains relative to the

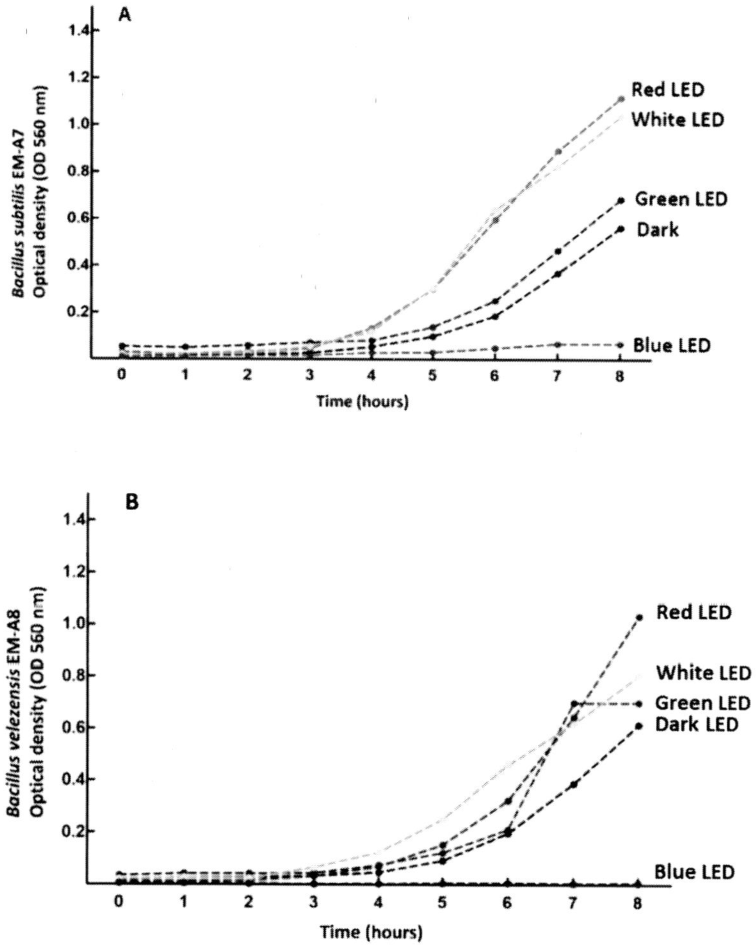

Figure 6. The effect of light on the growth of *Bacillus* strains. (A) Growth kinetics of the *Bacillus subtilis* EM-A7 during 8 h. (B) Growth kinetics of the *Bacillus velezensis* EM-A8 during 8 h of incubation at room temperature under different light sources. Each value represents the arithmetical mean of three independent experiments performed in duplicate.

3.3. The Effect of Light Quality on Ability to Inhibit Fungal Growth

The distance growth method was used to evaluate the interaction between the *Bacillus* strains and *Exserohilum turcicum* H4, which had been previously isolated from Aleppo sorghum in a field in Charras (Córdoba, Argentina). First, the bacterial strains and the fungus were grown by themselves for 7 days at room temperature, on potato dextrose agar (PDA) or MLA at Ψ = -1.38 and -4.09 MPa. During the incubation, the bacteria were exposed only to red or white light. These two qualities were selected based on previously reported results.

E. turcicum H4 was able to grow on its own on both solid media. The largest growth area was observed on PDA at Ψ= -4.09 MPa (982.56 ± 49.53 in mm^2) and the smallest ones on MLA at Ψ= -1.38 and -4.09 MPa (568.74 ± 5.77 and 585.39 ± 5.77, respectively). Both bacterial strains were able to develop on PDA and MLA at Ψ= -1.38 and -4.09 MPa in the absence of H4. EM-A8 developed the most on PDA at Ψ= -4.09 MPa under white light or in the dark (718.49 ± 38.91 and 812.07 ± 168.13, respectively). The biggest growth area for EM-A7 was recorded on PDA at Ψ= -1.38 MPa, under red or white light (1116.52 ± 179.66 and 1216.25 ± 307.27, respectively) (Figure 7).

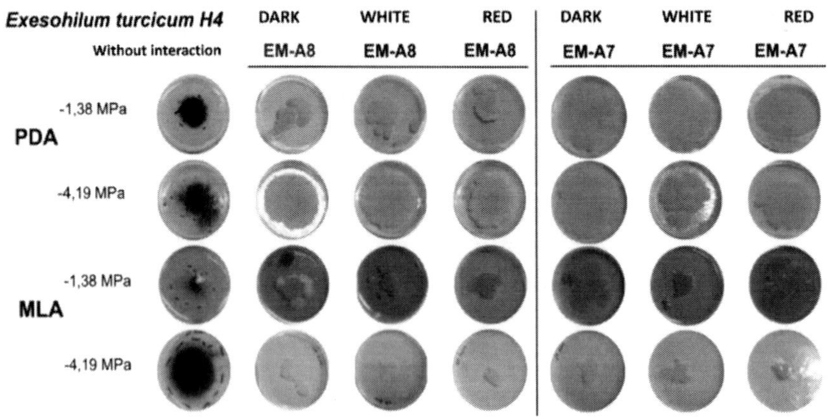

Figure 7. Growth of *Exserohilum turcicum* H4 strain without interaction with the bacterial strains EM-A7 and EM-A8. Growth of the *Bacillus subtilis* EM-A7 and *Bacillus velezensis* EM-A8 strains after exposure to two LED light sources, after being incubated at room temperature on PDA and MLA agar with modified water potential. PDA: potato dextrose agar; MLA: maize leaves agar; water potential values Ψ=-1.38, - 4.19 MPa.

After being incubated in the dark or under red or white light, each bacterial strain was grown on the same plate as the fungus, using the same media as in the previous step. The growth areas of the mycelial and bacterial cultures formed under these conditions were assessed on Image J 1.49v (Velez et al., 2018) and used to calculate the percentage of inhibition (PI%). Fungal growth was inhibited by the bacilli by between 50 and 74%. More specifically, on MLA at Ψ=-1.38 MPa, EM-A7 reduced fungal growth by 50, 72, and 71% after it had been respectively exposed to red light, white light or the dark. On the same medium, EM-A8 achieved 71% inhibition after being incubated under white light. On MLA at Ψ= -4.09 MPa, fungal growth dropped by 65-73% when the bacterial strains had been exposed to red light. In general, EM-A7 inhibited the growth of H4 on MLA at Ψ= -4.09 MPa and on PDA at both Ψ values when it had been incubated under red light. The same kind of light had this effect on EM-A8 only on PDA at Ψ= -4.09 MPa. EM-A8 was also able to reduce fungal growth on MLA and PDA at Ψ= -4.09 MPa after exposure to white light.

3.4. The Effect of Light Quality on the Relative Expression of Genes Associated with Biofilm and Motility

The relative expression of the genes *tasA*, *srfAA*, *blsA spo0A*, *epsA* and *hag*, associated with biofilm formation and motility, was evaluated in the two *Bacillus* strains after they were incubated for 8 h under white light, red light, or in the dark (control). The genes *rpsj* and *pky* were used as a reference for the normalization of relative mRNA levels in EM-A7 and EM-A8, respectively.

Figure 8 shows the relative expression scores for each gene. In EM-A7, the expression of *spo0A*, *epsA* and *hag* significantly increased with respect to the control ($p<0.05$) after exposure to red light and decreased significantly under white light. There were no changes in the expression of *tasA*, and a non-significant increase in that of *srfA* gene under both LED sources. In general, genes involved in biofilm formation and motility (*spo0A*, *epsA* and *hag*) were upregulated in this strain after exposure to red light.

In EM-A8, four of the genes tested showed statistically significant differences compared to the control ($p<0.05$). There was a significant increase in the relative expression of *tasA* and *epsA* under white light with respect to red light and the control. On the other hand, genes *spo0A* and *blsA* genes had a lower relative expression than the control after exposure to both LED sources, with significantly lower values in most cases. No significant differences with respect to the control were observed for *srfAA* and *hag*. Overall, genes involved in biofilm formation (*tasA* and *epsA*) were upregulated in this strain after exposure to white light, while gene *spo0A* gene, which regulates matrix gene expression and sporulation, was downregulated after exposure to both LED sources. In fact, the latter gene had the lowest expression, while *epsA* had the highest expression levels under the conditions tested.

Our results suggest a differential pattern of gene expression in strains EM-A7 and EM-A8 after 8 h of incubation in a liquid culture under different LED light sources.

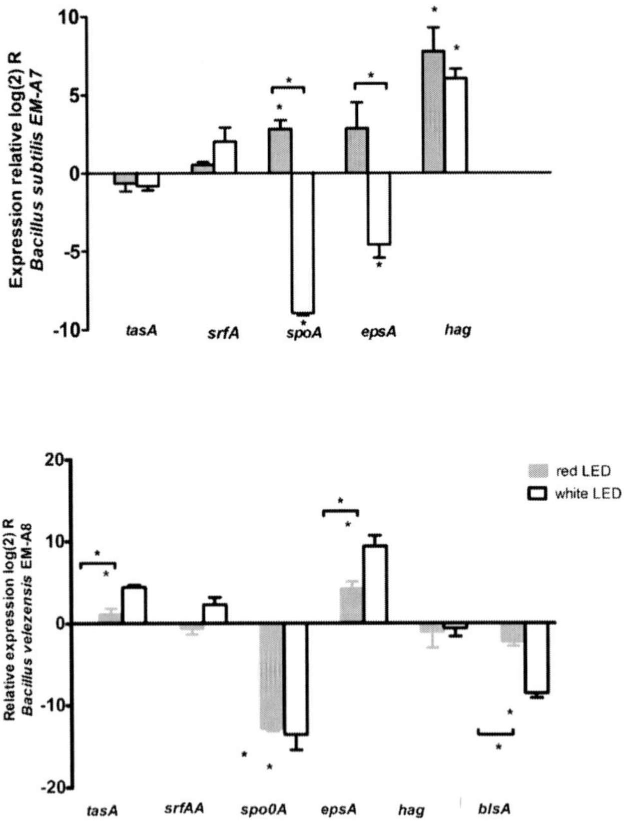

Figure 8. Fold change in mRNA levels of biofilm and motility genes of *Bacillus subtilis* EM-A7 and *Bacillus velezensis* EM-A8 incubated for 8 h under exposure to white and red LED light and compared with the control condition (darkness).

3.5. *In Silico* Search of Putative Photoreceptors and Light-Oxygen-Voltage (LOV)-Sensing Domain Sequences

Since the results obtained *in vitro* suggested that light quality is an influential factor on several aspects of EM-A7 and EM-A8, the draft genome of both strains (Fessia et al., 2024) were analyzed *in silico* to search for photoreceptor and LOV sequences. Amino acid sequences in the two strains, corresponding to putative photoreceptor and LOV sequences, were put through BLASTP so that they could be compared with reference sequences retrieved from NCBI with BLAST. The percentages of identity found are listed in Table 2.

In terms of photoreceptors, neither genome contained similar sequences to that of the multispecies photoactive yellow protein, or that of rhodopsin in *Kineococcus radiotolerans* SRS30216. Additionally, low identity percentages (30-20%) were observed when aligning proteins in the two bacilli with sequences corresponding to a photoreceptor in *Deinococcus radiodurans* R1 ATCC 13939 (GenBank accession number: AAF12261.1), and to the blue light- and temperature-regulated antirepressor BluF in *Escherichia coli* str. K-12 substr. MG1655 (GenBank accession number: NP_415681). In contrast, 98% (EM-A7) and 77% (EM-A8) identity were established with the sequence of the blue light GTP-binding receptor in *Bacillus subtilis subsp. subtilis str.* 168 (GenBank accession number: NP_390912).

When the draft genomes sequences of *Bacillus* strains were compared to LOV sequences, varying identity percentages (98-21%) were observed. In the case of strain EM-A7, we found a higher identity percentage for a putative protein kinase and blue-light FMN-binding sensor. These results suggest the presence of sequences potentially associated with putative photoreceptors associated with protein kinase and blue light sensor in both strains.

The biofilm produced on leaves by certain *Bacillus* spp. has also proven to play a role in controlling and preventing plant diseases (Arnaouteli et al. 2021; Fessia et al. 2022). Therefore, the design of consistent and effective biocontrol strategies against plant pathogens may depend, at least partly, on having thorough knowledge about how biofilm is influenced by changing environmental conditions. One of those conditions, light, is known to be implicated in the regulation of physiological processes in several microorganisms (Alsanius et al. 2021; Gharaie et al. 2017; Yu and Lee 2013), especially if these microorganisms live on the surface of the leaves. Therefore, analyzed phenotypic and genotypic responses to different light qualities as one step towards a more thorough comprehension of abiotic and biotic interactions in the phyllosphere.

In this chapter, we have summarized how light affects *B. subtilis* (EM-A7) and *B. velezensis* (EM-A8) growth and motility, as well as their ability to produce biofilm (complex colonies and pellicles) and antagonize a fungal pathogen. The sources of light to which the bacteria were exposed had different wavelengths and different photosynthetic flux densities. Fessia et al. (2022) demonstrated that the EM-A7 and EM-A8 were able to form pellicles and complex colonies under different environmental conditions (temperature, Ψ, nutrients). In Fessia et al. (2024) the same was described under different light sources.

Table 2. Known putative photoreceptors and LOV sequences. The percentages of identity correspond to homologous proteins in *Bacillus subtilis* EM-A7 and *Bacillus velezensis* EM-A8

GenBank accession number	Reference strain	Putative photoreceptor	% of identity EM-A7	% of identity EM-A8
NP_390912	*Bacillus subtilis subsp. subtilis str.* 168	blue light GTP-binding receptor	98%	77%
AAF12261.1	*Deinococcus radiodurans* R1 ATCC 13939 = DSM 20539	photoreceptor	30%	30%
NP_415681	*Escherichia coli str.* K-12 substr. MG1655	blue light- and temperature-regulated antirepressor BluF	28%	20%
WP_011235611.1	Multispecies	photoactive yellow protein	0%	0%
ZP_00618841.1	*Kineococcus radiotolerans* SRS30216	rhodopsin	0%	0%
GenBank accession number (protein)	Reference strain	LOV sequence	% of identity EM-A7	% of identity EM-A8
AAC00382.1	*Bacillus subtilis*	putative protein kinase	98%	77.0%
CAB15012.1	*Bacillus subtilis subsp. subtilis* str. 168	blue light FMN-binding sensor	98.47%	77.0%
AHF66268.1	*Pseudomonas cichorii* JBC1	histidine kinase	42%	42.0%
QED83298.1	*Pseudomonas amygdali pv. tabaci* str. ATCC 11528	PAS domain-containing sensor histidine kinase	36%	36.0%
AJY94712.1	*Xanthomonas citri* pv. citri	PAS domain S-box	38.21%	34.9%

QVK32491.1	*Pseudomonas syringae Susan*2139	PAS domain-containing protein	31.20%	30.0%
EFM0069695.1	*Escherichia coli*	blue light-responsive regulator BluF	29%	21.0%
AAM36973.1	*Xanthomonas citri* pv. citri str. 306	hypothetical protein XAC2120	0.0%	0.0%
AAM38122.1	*Xanthomonas citri pv. citri str.* 306	hypothetical protein XAC3278	0.0%	0.0%
AAO55826.1	*Pseudomonas syringae* pv. tomato str. DC3000	CHASE domain/PAS domain protein	0.0%	0.0%

Blue light significantly reduced the generation time (*g*) and the constant growth rate (*k*) in both study strains. Under red, green, and white light, these parameters were similar to the control, observing that *k* was significantly lower than the control when it was exposed to white light in EM-A7. In EM-A8, red light enhanced growth with respect to the control, while green and white light led to values similar to the control. Exposure to light has been associated with oxidative stress and toxicity in bacteria (Río-Álvarez et al. 2014). While red, green, and white light were evidently not harmful for our *Bacillus* strains, further studies should investigate whether the reduction in growth recorded under blue light is indeed related to phenomena such as oxidative stress and toxicity. Yu and Lee (2013) found that biofilm production by *B. amyloliquefaciens* JBC36 was significantly enhanced under red light at 240 μmol m-2 s-1. They also described thicker colonies being formed on swarming plates upon exposure to this red light, and suggested that the increase in thickness was ascribable to an increase in biofilm formation. Alsanius et al. (2021) noted that the amount of biofilm produced by *Pseudomonas* strains changed in response to nutrient availability and exposure to light. Light quality also influenced biofilm formation by three commercial BCAs (Karlsson et al. 2023). In the case of *B. amyloliquefaciens,* it was higher in the dark and under red light. The same authors reported that specific combinations of light quality (green or blue) and carbon source (sucrose, fructose, dulcitol) increased biofilm formation, growth and respiration.

The strains EM-A7 and EM-A8 were able to significantly reduce the growth of *E. turcicum* H4 after being exposed to red and white light as control conditions (dark). When studying the activity of JBC36 against the mycelial growth of *Penicillium digitatum* under different light wavelengths, Yu and Lee (2013) similarly found that the average size of the inhibition zones was bigger under red and green light than in the dark. Karlsson et al. (2023) have indicated that declining numbers of introduced biocontrol agents can be a problem when they are applied to seeds, roots, and leaves. This might be related to the chosen application techniques or to poor biocontrol agent adaptation to the commercial growing environment. For this reason, the seeming relevance of light wavelength for biofilm formation, bacterial survival, and antipathogenic activity on the leaf should be corroborated by larger-scale studies.

Genomic analyses would also make it possible to ascertain whether bacterial photoreceptor proteins are involved in the responses to light observed here. Light-sensitive receptor proteins in bacteria, such as light oxygen voltage receptor protein (LOV-HK), blue light-sensing protein using FAD (BLUF), and phytochrome, have been reported in *Pseudomonas aeruginosa*, *P. syringae*, *Xanthomonas axonopodis* pv. citri, *Rhodopseudomonas pallustris,* and *Bacillus subtilis*, respectively. We demonstrated that growth rate, the formation of complex colonies and pellicles, biofilm quantification, motility, and antagonism towards *E. turcicum* of *B. subtilis* EM-A7 and *B. velezensis* EM-A8 were influenced by light quality. In general, the effects of blue light were negative or inhibitory, and those of red and white light were overall positive (Fessia et al., 2024), in agreement with other studies (Alsanius et al., 2021; Gharaie et al., 2017; Karlsson et al., 2023; Río-Álvarez et al., 2014; Yu & Lee, 2013). The LOV and BLUF are associated with blue light sensing in plants and microorganisms, and they are functions in attachment, multicellularity, production of adhesion proteins, and virulence (Alsanius et al., 2019). In the present study, we performed an *in-silico* analysis to detect sequences conserved in the genomes of *Bacillus* strains associated with photoreceptors and LOV sequences, using reference sequences obtained from GenBank. Proteins with light, oxygen, or voltage (LOV) domains have been associated with different cellular responses such as virulence, gene expression, stress responses, photosynthesis pigment synthesis, motility, and attachment of pathogenic bacteria, as well as growth (Bonomi et al., 2012; Herrou & Crosson, 2011; Möglich et al., 2010; Río-Álvarez et al., 2014). Photo-sensing in bacteria can trigger a cascade of reactions within the cell. Signaling molecules and regulatory proteins can lead to changes in gene regulation, which may modify the behavior and lifestyle of the bacteria involved (Alsanius et al., 2024). LOV, PYP, and BLUF have been shown under different light conditions, which are shown to control the transition between planktonic single-cell lifestyle into a sessile multicellular lifestyle (Alsanius et al., 2021; Karlsson et al., 2023).

Conclusion

All physiological and genetic aspects evaluated in two biocontrol agents, EM-A7 and EM-A8, were influenced by light quality, as well as other environmental factors. The changes were wavelength-dependent, that is to say, the effects of red and white light were generally positive. Since all these factors (particularly biofilm formation and motility) can be important for the establishment and survival of *Bacillus* strains on leaves, as well as for their efficacy against pathogens, light could be an important element to consider in the design of biocontrol strategies.

References

Adivitiya, B., & Khasa, Y. P. (2017). Microbes as biocontrol agents. In *Probiotics and Plant Health* (pp. 507–552). Springer, Singapore. https://doi.org/10.1007/978-981-10-3473-2_24.

Alabouvette, C., Heilig, U., & Cordier, C. (2012). Microbial control of plant diseases. *Beneficial Microorganisms in Agriculture, Food and the Environment: Safety Assessment and Regulation*, 96–111. https://doi.org/10.1079/9781845938109.0096.

Alsanius, B. W., Hellström, M., Bergstrand, K.-J., Vetukuri, R., Becher, P., & Karlsson, M. E. (2024). The power of light from a non-phototrophic perspective: a phyllosphere dilemma. *Frontiers in Photobiology*, *2*, 1432066. https://doi.org/10.3389/FPHBI.2024.1432066.

Alsanius, B. W., Karlsson, M., Rosberg, A. K., Dorais, M., Naznin, M. T., Khalil, S., & Bergstrand, K. J. (2019). Light and microbial lifestyle: The impact of light quality on plant–microbe interactions in horticultural production systems—A review. *Horticulturae*, *5*(2), 1–24. https://doi.org/10.3390/horticulturae5020041.

Alsanius, B. W., Vaas, L., Gharaie, S., Karlsson, M. E., Rosberg, A. K., Wohanka, W., Khalil, S., & Windstam, S. (2021). Dining in Blue Light Impairs the Appetite of Some Leaf Epiphytes. *Frontiers in Microbiology*, *12*, 725021. https://doi.org/10.3389/FMICB.2021.725021/BIBTEX.

Arnaouteli, S., Bamford, N. C., Stanley-Wall, N. R., & Kovács, Á. T. (2021). *Bacillus subtilis* biofilm formation and social interactions. *Nature Reviews Microbiology* 19:600–614. https://doi.org/10.1038/s41579-021-00540-9.

Bartolini, M., & Grau, R. (2019). Assessing different ways of Bacillus subtilis spreading over abiotic surfaces. *Bio-Protocol* 9:1–10. https://doi.org/10.21769/bioprotoc.3425.

Bashir, I., War, A. F., Rafiq, I., Reshi, Z. A., Rashid, I., & Shouche, Y. S. (2022). Phyllosphere microbiome: Diversity and functions. *Microbiological Research*, *254*, 126888. https://doi.org/10.1016/J.MICRES.2021.126888.

Berger, D. K. (2024). Diseases of maize/corn. *Agrios' Plant Pathology, Sixth Edition*, 739–746. https://doi.org/10.1016/B978-0-12-822429-8.00030-3.

Bonomi, H. R., Posadas, D. M., Paris, G., Del Carmen Carrica, M., Frederickson, M., Pietrasanta, L. I., Bogomolni, R. A., Zorreguieta, A., & Goldbaum, F. A. (2012). Light regulates attachment, exopolysaccharide production, and nodulation in *Rhizobium leguminosarum* through an LOV-histidine kinase photoreceptor. *Proceedings of the National Academy of Sciences of the United States of America*, *109*(30), 12135–12140. https://doi.org/10.1073/PNAS.1121292109.

Carmona, M., & Sautua, F. (2016). *Enfermedades del maíz y Umbrales para su control 2015-2016*. https://www.agroconsultasonline.com.ar//documento.html?op=v& documento_id=658ten.

Couretot, L., Parisi, L., & Ferrari, L. (2012). Efecto de fungicidas foliares y momento de aplicación sobre la severidad de tizón foliar y enfermedades de raíz y tallo en maíz. In *XIV Jornadas Fitosanitarias Argentina*. https://inta.gob.ar/sites/default/files/script-tmp-inta_pergamino_efecto_de_fungicidas_foliares_y_moment.pdf.

Darino, M. A., Rochi, L., Lia, V. V., Kreff, E. D., Pergolesi, M. F., Ingala, L. R., Dieguez, M. J., & Sacco, F. (2018). Virulence characterization and identification of resistant maize lines to *Puccinia sorghi* Schwein present in the Argentine Corn Belt region. *Plant Disease*, *100*(4), 770–776. https://doi.org/10.1094/PDIS-06-15-0639-RE.

De Rossi, R. L. (2017). Situación fitosanitaria del cultivo de maíz en Argentina. *IV° Congreso Argentino de Fitopatología*. https://www.researchgate.net/publication/322070791_SITUACION_FITOSANITARIA_DEL_CULTIVO_DE_MAIZ_EN_ARGENTINA.

Devi, B., Gupta, S. K., Singh, G. & Prasad, P. (2020). Efficacy of new generation fungicides against French bean rust caused by Uromyces appendiculatus. *Phytoparasitica* 48(4):535– 43. https://doi.org/10.1007/s12600-020-00820-9.

Estimaciones | Bolsa de Comercio de Rosario. (n.d.). Retrieved March 25, 2025, from https://www.bcr.com.ar/es/mercados/gea/estimaciones-nacionales-de-produccion/estimaciones.

Fessia, A., Ponzio, R., Arcibia, L., Barros, G., & Nesci, A. (2024). Effects of different light wavelengths on *Bacillus subtilis* and *Bacillus velezensis*, two biocontrol agents isolated from the maize phyllosphere. *Archives of Microbiology, 206:104*. https://doi.org/10.1007/s00203-024-03836-5.

Fessia, A., Sartori, M., García, D., Fernández, L., Ponzio, R., Barros, G., & Nesci, A. (2022). *In vitro* studies of biofilm-forming *Bacillus s*trains, biocontrol agents isolated from the maize phyllosphere. *Biofilm*, *4*, 100097. https://doi.org/10.1016/J.BIOFLM.2022.100097.

Fessia, A., Sartori, M., Orlando, J., Barros, G., & Nesci, A. (2024). Draft genome sequences of two biocontrol agents isolated from the maize phyllosphere: *Bacillus subtilis* strain EM-A7 and *Bacillus velezensis* strain EM-A8. *Heliyon*, *10*(12). https://doi.org/10.1016/J.HELIYON.2024.E32607.

Formento, N. A. (2010). Enfermedades foliares reemergentes del cultivo de maíz: (*Puccinia sorghi* y *Puccinia polysora*), tizón foliar (*Exserohilum turcicum*) y mancha ocular (*Kabatiella zeae*). *Estacion Experimental Agropecuaria Paraná*, 89–100.

FRAC. (2016). *Mode of action of fungicides. Fungicide Resistance Action Committee.*

Garg, R., & Smith, C. J. (2014). Predicting the bioconcentration factor of highly hydrophobic organic chemicals. *Food and Chemical Toxicology*, *69*, 252–259. https://doi.org/10.1016/j.fct.2014.03.035.

Gharaie, S., Vaas, L. A. I., Rosberg, A. K., Windstam, S. T., Karlsson, M. E., Bergstrand, K. J., Khalil, S., Wohanka, W., & Alsanius, B. W. (2017). Light spectrum modifies the utilization pattern of energy sources in *Pseudomonas* sp. DR 5-09. *PLoS ONE*, *12*(12), 1–24. https://doi.org/10.1371/journal.pone.0189862.

Gomelsky, M., & Hoff, W. D. (2011). Light helps bacteria make important lifestyle decisions. *Trends in Microbiology*, *19*(9), 441–448. https://doi.org/10.1016/J.TIM.2011.05.002.

Hashem, A., Tabassum, B., & Fathi Abd Allah, E. (2019). *Bacillus subtilis*: A plant-growth-promoting rhizobacterium that also impacts biotic stress. In *Saudi Journal of Biological Sciences* 26 (6) pp. 1291–1297. Elsevier. https://doi.org/10.1016/j.sjbs.2019.05.004.

Herrou, J., & Crosson, S. (2011). Function, structure and mechanism of bacterial photosensory LOV proteins. *Nature Reviews. Microbiology*, *9*(10), 713–723. https://doi.org/10.1038/NRMICRO2622.

Hill, S. B. (1998). Redesigning agroecosystems for environmental sustainability: A deep systems approach. *Systems Research and Behavioral Science*, *15*(5), 391–402. https://doi.org/10.1002/(SICI)1099-1743(1998090)15:5<391::AID-SRES266>3.0.CO;2-0.

Jat, H. S., Datta, A., Choudhary, M., Sharma, P. C., & Jat, M. L. (2021). Conservation Agriculture: factors and drivers of adoption and scalable innovative practices in Indo-Gangetic plains of India– a review. *International Journal of Agricultural Sustainability*, *19*(1), 40–55. https://doi.org/10.1080/14735903.2020.1817655.

Jindal, K., Zhu, X., Woldemariam, T., Tenuta, A., & Jindal, M. (2019). *Maize inbreds for multiple resistance breeding against major foliar, ear and stalk rot diseases.* https://core.ac.uk/download/pdf/230660718.pdf.

Karlsson, M. E., Hellström, M., Flöhr, A., Bergstrand, K. J., & Alsanius, B. W. (2023). The power of light: Impact on the performance of biocontrol agents under minimal nutrient conditions. *Frontiers in Microbiology*, *14*, 1087639. https://doi.org/10.3389/FMICB.2023.1087639/BIBTEX.

Karp, D. S., Gennet, S., Kilonzo, C., Partyka, M., Chaumont, N., Atwill, E. R., & Kremen, C. (2015). Comanaging fresh produce for nature conservation and food safety. *Proceedings of the National Academy of Sciences of the United States of America*, *112*(35), 11126–11131. https://doi.org/10.1073/pnas.1508435112.

Knoema (2023). World: Maize production quantity. Available online at: https://knoema.com/atlas/World/topics/Agriculture/Crops-Production-Quantity-tonnes/Maize-production.

Kotze, R. G., van der Merwe, C. F., Crampton, B. G., & Kritzinger, Q. (2019). A histological assessment of the infection strategy of *Exserohilum turcicum* in maize. *Plant Pathology*, *68*(3), 504–512. https://doi.org/10.1111/PPA.12961.

Lahlali, R., Ezrari, S., Radouane, N., Kenfaoui, J., Esmaeel, Q., El Hamss, H., Belabess, Z., & Barka, E. A. (2022). Biological Control of Plant Pathogens: A Global

Perspective. *Microorganisms*, *10*(3), 596. https://doi.org/10.3390/microorganisms 10030596.

Leonard, K. (1974). *Bipolaris maydis* race and mating type frequencies in North Carolina. *Plant Disease Report* 58, 529-531. https://www.cabidigitallibrary.org/doi/full/10.5555/19741314054.

Möglich, A., Yang, X., Ayers, R. A., & Moffat, K. (2010). Structure and function of plant photoreceptors. *Annual Review of Plant Biology*, *61*, 21–47. https://doi.org/10.1146/ANNUREV-ARPLANT-042809-112259.

Morris, C. E. (2002) Phyllosphere. In: *Encyclopedia of life sciences*. London: Nature Publishing Group.

Oberpichler, I., Rosen, R., Rasouly, A., Vugman, M., Ron, E. Z., & Lamparter, T. (2008). Light affects motility and infectivity of *Agrobacterium tumefaciens. Environmental Microbiology*, *10*(8). https://doi.org/10.1111/j.1462-2920.2008.01618.x.

Ray, D. K., Mueller, N. D., West, P. C., & Foley, J. A. (2013). Yield trends are insufficient to double global crop production by 2050. *Plos One* 8, e66428. https://doi.org//10.1371/journal.pone.0066428.

Reis, M. G. dos, & Ribeiro, A. (2020). Conversion factors and general equations applied in agricultural and forest meteorology. *Agrometeoros*, *27*(2). https://doi.org/10.31062/AGROM.V27I2.26527.

Río-Álvarez, I., Rodríguez-Herva, J. J., Martínez, P. M., González-Melendi, P., García-Casado, G., Rodríguez-Palenzuela, P., & López-Solanilla, E. (2014). Light regulates motility, attachment and virulence in the plant pathogen *Pseudomonas syringae* pv tomato DC3000. *Environmental Microbiology*, *16*(7), 2072–2085. https://doi.org/10.1111/1462-2920.12240.

Röös, E., Bajželj, B., Smith, P., Patel, M., Little, D., & Garnett, T. (2017). Greedy or needy? Land use and climate impacts of food in 2050 under different livestock futures. *Global Environmental Change*, *47*, 1–12. https://doi.org/10.1016/J.GLOENVCHA.2017.09.001.

Sartori, M., Bonacci, M., Barra, P., Fessia, A., Etcheverry, M., Nesci, A., & Barros, G. (2020). Studies on possible modes of action and tolerance to environmental stress conditions of different biocontrol agents of foliar diseases in maize. *Agricultural Sciences*, *11*(06), 552–566. https://doi.org/10.4236/as.2020.116035.

Sartori, M., Nesci, A., Formento, Á., & Etcheverry, M. (2015). Selection of potential biological control of *Exserohilum turcicum* with epiphytic microorganisms from maize. *Revista Argentina de Microbiología*, *47*(1), 62–71. https://doi.org/10.1016/j.ram.2015.01.002.

Sartori, M., Nesci, A., García, J., Passone, M. A., Montemarani, A., & Etcheverry, M. (2017). Efficacy of epiphytic bacteria to prevent northern leaf blight caused by *Exserohilum turcicum* in maize. *Revista Argentina de Microbiologia*, *49*(1), 75–82. https://doi.org/10.1016/j.ram.2016.09.008.

Savary, S., Willocquet, L., Pethybridge, S. J., Esker, P., Mcroberts, N., & Nelson, A. (2019). The global burden of pathogens and pests on major food crops. *Nature Ecology & Evolution,* https://doi.org/10.1038/s41559-018-0793-y.

Singh, B. K., Trivedi, P., Egidi, E., Macdonald, C. A., & Delgado-Baquerizo, M. (2020). Crop microbiome and sustainable agriculture. *Nature Reviews Microbiology 2020 18:11*, *18*(11), 601–602. https://doi.org/10.1038/s41579-020-00446-y.

Sivakumar, N., Sathishkumar, R., Selvakumar, G., Shyamkumar, R., & Arjunekumar, K. (2020). Phyllospheric microbiomes: diversity, ecological significance, and biotechnological applications. In *Plant Microbiomes for Sustainable Agriculture* 5 (25), pp. 113–172. Nature Publishing Group. https://doi.org/10.1007/978-3-030-38453-1_5.

Smith, D. & Proost, R. (2020) Fungicide resistance management in corn, soybean, and wheat in Wisconsin (A3878). This publication is available from your Wisconsin county Extension office (counties.extension.wisc.edu).

Stenberg, J. A., Sundh, I., Becher, P. G., Björkman, C., Dubey, M., Egan, P. A. et al. (2021) When is it biological control? A framework of definitions, mechanisms, and classifications. *Journal of Pest Science*, 94, 665–676.

Tandzi, N. L., & Mutengwa, S. C. (2020). Factors Affecting Yield of Crops. *Agronomy - Climate Change and Food Security*. https://doi.org/10.5772/INTECHOPEN.90672.

Thapa, S., & Prasanna, R. (2018). Prospecting the characteristics and significance of the phyllosphere microbiome. In *Annals of Microbiology* 68 (5) 229–245. Annals of Microbiology. https://doi.org/10.1007/s13213-018-1331-5.

Van Der Horst, M. A., & Hellingwerf, K. J. (2004). Photoreceptor proteins, "star actors of modern times": a review of the functional dynamics in the structure of representative members of six different photoreceptor families. *Accounts of Chemical Research*, *37*(1), 13–20. https://doi.org/10.1021/AR020219D.

Velez, P., Espinosa-Asuar, L., Figueroa, M., Gasca-Pineda, J., Aguirre-von-Wobeser, E., Eguiarte, L. E., Hernandez-Monroy, A., & Souza, V. (2018). Nutrient-dependent cross-kingdom interactions: Fungi and bacteria from an oligotrophic desert oasis. *Frontiers in Microbiology*, *9*(AUG), 358832. https://doi.org/10.3389/FMICB.2018.01755/BIBTEX.

Vorholt, J. A. (2012). Microbial life in the phyllosphere. In *Nature Reviews Microbiology* 10 (12), pp. 828–840. Nat Rev Microbiol. https://doi.org/10.1038/nrmicro2910.

White, D. (1999). *Compendium of corn diseases*. https://www.cabidigitallibrary.org/doi/full/10.5555/19991701011.

Wu, L., McGrane, R. S., & Beattie, G. A. (2013). Light regulation of swarming motility in *Pseudomonas syringae* integrates signaling pathways mediated by a bacteriophytochrome and a LOV protein. *MBio*, *4*(3). https://doi.org/10.1128/mBio.00334-13.

Yu, S. M., & Lee, Y. H. (2013). Effect of light quality on *Bacillus amyloliquefaciens* JBC36 and its biocontrol efficacy. *Biological Control*, *64*(3), 203–210. https://doi.org/10.1016/j.biocontrol.2012.11.004.

Chapter 4

Yeasts as Biocontrol Agents: Potential Applications in Wine-Making and Fruits Conservation

Miguel Fernández de Ullivarri[1]
Emilse V. Bentencourt[2]
and Lucía M. Mendoza[2,*]
[1]APC Microbiome Ireland, University College Cork, Cork, Ireland
[2]Centro de Referencia para Lactobacilos (CERELA), Tucumán, Argentina

Abstract

Food contamination is a major concern for food production, quality, and safety worldwide. To address this, various antimicrobial agents are employed to inhibit spoilage microorganisms, with chemical additives being the most widely used. However, in recent years, there has been a growing tendency to reduce or replace their use. In this context, yeasts emerge as promising biocontrol agents capable of inhibiting a wide range of spoilage microorganisms, including yeasts, fungi, and bacteria. Among these are killer yeasts, which produce proteins or glycoproteins known as killer toxins. Additional inhibitory mechanisms include competition and the production of bioactive compounds (e.g., acids, esters, alcohols).

In winemaking, certain strains of yeasts and bacteria can compromise wine quality by producing undesirable metabolites, such as biogenic amines and volatile phenol compounds, while fungi can produce mycotoxins. Sulfur dioxide is commonly applied to grapes to inhibit

---

* Corresponding Author's Email: lmendoza@cerela.org.ar

In: Biocontrol Agents (BCAs)
Editor: Daiana García
ISBN: 979-8-90134-102-5

spoilage microorganisms. However, lower concentrations are now being used, prompting the exploration of alternative strategies. Killer or antagonistic yeasts represent one of the most promising alternatives.
Similarly, microbial contamination also affects fruits such as grapes, raisins, blueberries, and strawberries, particularly during post-harvest and storage stages. Fungal contamination is the most prevalent issue, with some yeast strains demonstrating strong inhibitory activity against various species of spoilage fungi and yeasts.
Yeasts as biocontrol agents represent a promising and eco-friendly alternative to chemical preservatives, offering effective control over a wide spectrum of spoilage microorganisms that contaminate beverages like wine as well as a variety of fruits.

Keywords: yeasts, biocontrol, spoilage, food, wine

1. Introduction

Microbiological contamination of food and beverages remains a major concern for food safety and public health. Spoilage microorganisms such as bacteria, yeasts, and filamentous fungi not only deteriorate product quality but can also pose significant health risks by producing harmful metabolites, including mycotoxins and biogenic amines [1-3]. As global demand for minimally processed and preservative-free products increases, the food industry faces the challenge of ensuring safety while minimizing the use of synthetic chemical preservatives [4, 5].

Chemical antimicrobials such as sulfur dioxide, sorbates, and benzoates have been widely used to control microbial growth in food and beverage matrices [6]. However, their long-term safety, impact on sensory attributes, and consumer rejection due to perceived health risks have motivated the search for alternative, more sustainable strategies [7-9]. In this context, biological control methods, particularly those involving microorganisms with antagonistic activity, have gained attention as promising tools for food preservation [10, 11].

Among biocontrol agents, yeasts have emerged as effective antagonists against a broad spectrum of spoilage microorganisms. Several yeast species naturally colonize fruits, musts, and fermented products, exhibiting mechanisms of microbial inhibition that include nutrient competition, secretion of volatile organic compounds, acidification, and production of killer toxins, proteins or glycoproteins capable of selectively killing susceptible microorganisms [12-15]. Unlike chemical agents, antagonistic yeasts offer targeted inhibition without altering food quality or posing toxicological concerns [16].

Killer yeasts, such as *Saccharomyces cerevisiae*, *Wickerhamomyces anomalus*, *Metschnikowia pulcherrima*, and *Pichia* spp., are particularly promising due to their ability to secrete killer toxins with activity against a range of spoilage yeasts, molds, and even bacteria [17-19]. Their effectiveness has been demonstrated in various food matrices, including grapes, lemons, wine, and fermented beverages, often outperforming conventional preservatives under certain conditions [20-22].

Furthermore, the use of antagonistic yeasts aligns with current regulatory and market trends favoring natural, clean-label products [23]. Yeast-based biocontrol strategies have been studied not only in fresh produce but also in post-harvest management, storage, and during fermentation, indicating their versatility and practical relevance for different stages of food processing [13, 17, 24, 25].

Despite the increasing body of research supporting their efficacy, several challenges remain in the implementation of antagonistic yeasts as standard biocontrol agents. These include variability in strain performance under industrial conditions, limited understanding of toxin biosynthesis pathways, and the need for regulatory frameworks specific to microbial bio-preservatives [26, 27]. Nonetheless, continued advances in microbial ecology, genomics, and fermentation technology are expected to enhance the selection and application of effective yeast strains in food systems [28].

2. Food Contamination

Microbial contamination significantly compromises the safety, quality, and shelf life of food and beverage products. Contaminants include a wide variety of bacteria, yeasts, and molds that can proliferate at various stages of production, processing, storage, and distribution [1-3]. These microorganisms not only cause spoilage but also produce harmful metabolites, such as mycotoxins, biogenic amines, and ethyl carbamate, which pose serious health risks to consumers [29-31].

In fruit-based products, contamination frequently occurs during pre-harvest, harvest, and post-harvest handling. Fungal species such as *Penicillium*, *Aspergillus*, and *Botrytis cinerea* are prevalent on grapes, strawberries, and other berries, leading to decay and potential mycotoxin production [32-34]. *B. cinerea*, responsible for gray mold, is one of the most economically damaging post-harvest pathogens in fresh fruits [35]. *Aspergillus carbonarius* and *A. niger* are notable for their production of ochratoxin A (OTA), particularly in wine grapes and raisins [36].

In fermented beverages such as wine, microbial stability is critical. Spoilage yeasts like *Brettanomyces bruxellensis* can grow during or after fermentation, producing undesirable volatile phenols (e.g., 4-ethylphenol, 4-ethylguaiacol) that impart off-flavors and reduce product quality [37, 38]. Certain lactic acid bacteria, including *Lactobacillus* and *Pediococcus*, can produce excessive acetic acid and biogenic amines such as histamine and tyramine, posing potential allergenic and toxicological effects [39-41].

Soft drinks, fruit juices, and other non-alcoholic beverages are also vulnerable to microbial spoilage, particularly by osmotolerant and acidophilic yeasts such as *Zygosaccharomyces bailii* and *S. cerevisiae*, which can survive in high-sugar, low-pH environments [42, 43]. These yeasts cause gas production, turbidity, and off-flavors, leading to product rejection [44].

Contamination may also arise from biofilms formed on equipment and surfaces. These biofilms, composed of microbial cells embedded in extracellular polymeric substances, are more resistant to sanitizers and preservatives [45, 46]. As a result, they cause recurrent contamination and complicate industrial sanitation.

Additionally, poor handling practices, inadequate hygiene, and suboptimal storage conditions further exacerbate contamination. Psychrotrophic spoilage organisms such as *Pseudomonas* spp. can proliferate in refrigerated products degrading proteins and lipids in dairy and meat products [47]. In cereal and nut products, xerophilic molds like *Aspergillus flavus* and *Penicillium verrucosum* are common sources of aflatoxins and ochratoxins, particularly in warm, humid environments [48, 49].

The diversity and adaptability of spoilage and pathogenic microorganisms highlight the need for effective and specific control measures. Standard chemical preservatives are often used but may be insufficient under certain stress conditions or in minimally processed products, which have fewer intrinsic barriers to microbial growth [50, 51]. Understanding the ecology, physiology, and spoilage mechanisms of contaminant organisms is essential for the development of targeted and sustainable preservation strategies [52, 53].

3. Winemaking Process: Production, Contamination and Strategies to Control Spoilage Microorganisms

The winemaking process consists of a series of operations carried out from grape harvesting to wine production. This process involves physical, chemical, and biological transformations, with microbially driven fermentations playing a central role. Yeasts, primarily *Saccharomyces cerevisiae*, convert grape must sugars into ethanol and carbon dioxide. During the alcoholic fermentation (AF), various compounds are also produced by yeasts, and they influence on the wine aroma and flavour. In certain types of wines, lactic acid bacteria (LAB) such as *Oenococcus oeni* may conduct a secondary fermentation known as malolactic fermentation (MLF), which reduces acidity and softens the sensory profile of the final product [54-57].

Wine production is a complex biotechnological process that can vary depending on the region and the specific type of wine being produced. Grapes are harvested when they reach optimal ripeness for achieving the desired wine style. The quality of the wine largely depends on the degree of grape ripeness, as the content of sugars, acids, and aromatic compounds significantly influences the fermentation process [55]. The grapes are then crushed to break the skins and release the must. The fruit is also determined to remove the stems, which may otherwise impart excessively bitter or herbaceous characteristics to the wine. At this point, sulfur dioxide (SO_2) is commonly added for the first time to control the development of undesirable microorganisms and limit must oxidation (Figure 1). The sequence of subsequent steps depends on the type of wine. Wines are classified by color into red, rosé, and white. Red wines are produced from red grape varieties, with color extracted from skins during extended contact with the must. Rosé wines undergo shorter skin contact. White wines are made from the must of either white or red grapes, but in the latter case, skins must be removed promptly to avoid pigmentation. For red wines, AF occurs while the must remain in contact with the skins. As fermentation progresses, the skins rise to form a "cap" on the liquid surface [54].

AF is the central process in winemaking. It is a metabolic pathway carried out by *S. cerevisiae* and other yeasts. In the early stages of AF, various non-*Saccharomyces* yeasts such as *Hanseniaspora*, *Candida*, *Pichia*, and *Metschnikowia* play an active role due to their rapid proliferation and natural presence on the grape surface. However, their low ethanol tolerance and limited fermentative capacity result in their displacement by *S. cerevisiae*, which completes the process due to its high alcohol resistance and efficiency in sugar conversion [55, 56].

MLF typically follows AF and is an enzymatic decarboxylation of L-malic acid to L-lactic acid with carbon dioxide release [57]. This transformation is primarily carried out by *O. oeni*, a LAB species highly adapted to wine's harsh conditions: high ethanol, low pH, elevated SO_2, and low temperatures. Other LAB, such as *Lactobacillus* and *Pediococcus*, may also participate [58]. From an oenological perspective, MLF contributes to wine deacidification by reducing total acidity and raising the pH by approximately 0.2 to 0.5 units. It also enhances microbial stability, as the consumption of malic acid reduces the availability of this carbon source for undesirable microorganisms. Moreover, MLF improves the wine's flavor profile through the action of LAB enzymes, which produce secondary metabolites such as diacetyl, esters, and alcohols [59].

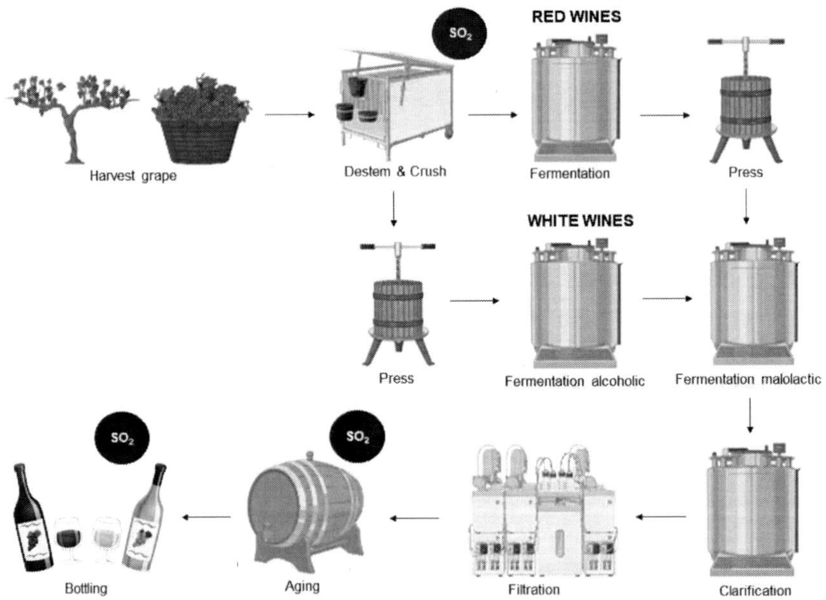

Figure 1. The main steps in wine production. The black circles indicate the main phases where SO_2 is generally added to avoid microbial contamination.

Both AF and MLF can occur spontaneously due to native microorganisms found on grape surfaces and within the winery environment. These microbial populations are heterogeneous and can vary significantly even among wineries located within the same region [60, 61]. The use of this indigenous microbiota is considered to contribute to the uniqueness of the wine, reflecting specific characteristics of each viticultural region (*"terroir"*) [62]. However, the natural variability of these microbial communities introduces a degree of unpredictability in the sensory attributes of the wine, resulting in a lack of reproducibility in final quality, even between vintages from the same winery [63].

Microbial variability increases the risk of contamination by undesirable microorganisms, thereby necessitating greater use of antimicrobial agents. In recent years, oenologists have increasingly recognized the benefits of inoculating musts with starter cultures, comprising both yeasts and LAB, at defined concentrations to exert greater control over the timing and progression of fermentations. This practice reduces the formation of undesirable by-products, accelerates fermentation, and minimizes the risk of contamination or fermentation arrest. Furthermore, by enabling more precise microbiological control, it facilitates the production of wines with more consistent quality across different vintages [64].

In the final stages, wine undergoes clarification, stabilization, and filtration (Figure 1). Clarification involves the removal of suspended particles through the addition of fining agents, which aggregate and sediment unwanted solids. Stabilization aims to ensure the physicochemical and microbiological stability of the wine, preventing precipitation or undesired fermentations during ageing. Filtration refines the clarity of the wine and ensures its microbiological stability by retaining microorganisms and residual matter.

Ageing is the subsequent process, by which wine is left to mature over a specific period, either in stainless steel tanks, wooden barrels, or bottles. The objective of ageing is to develop favorable sensory attributes, stabilize color, and promote the polymerization of phenolic compounds such as tannins, thereby reducing astringency and enhancing wine structure. The duration and type of ageing vessel strongly influence the wine's profile. Depending on the length of the ageing period, SO_2 is commonly added to prevent oxidation and maintain microbiological stability, thus avoiding microbial spoilage in barrels.

Bottling is the final step. At this stage, the wine is microbiologically and chemically stable and ready for preservation. Final SO_2 adjustments may be made to ensure post-bottling stability. Once bottled, the wine may be released to the market or continue to evolve within the bottle, depending on the intended style [65].

Various issues may compromise wine quality, stability, and safety. One of the most significant is microbial contamination. Spoilage yeast species such as *B. bruxellensis*, can produce volatile phenols that impart off-flavors described as "horse sweat" or "animal leather" [66]. In addition, this yeast species can synthesize biogenic amines and acetic acid [67].

Acetic acid bacteria, such as *Acetobacter aceti*, *Acetobacter pasteurianus*, and *Gluconobacter oxydans*, may also proliferate, oxidizing ethanol into acetic acid, which results in a vinegary off-flavor in the wine [68]. Additionally, contaminant filamentous fungi (*Aspergillus* and *Penicillium*) may be present on grapes while still in the vineyard. These fungi can produce mycotoxins such as ochratoxin A, a toxic and potentially carcinogenic compound that may persist in finished wine [69].

Furthermore, certain LAB can produce compounds that alter the wine's organoleptic properties, such as acetic acid and volatile phenols, or compromise its hygienic quality through the formation of substances with potential adverse health effects, including ethyl carbamate and biogenic amines [70].

To manage these risks, the use of SO_2 is routinely employed in winemaking. This compound, the main additive used in the wine industry, acts as a preservative primarily through its antimicrobial and antioxidant properties. It can be applied at various stages of the process, from harvest to bottling (Figure 1), and its effectiveness allows the winemaker to exert greater control during wine production, enhancing the microbiological and sensory stability of the final product.

3.1. Sulphur Dioxide Use in Winemaking

Sulphur dioxide (SO_2) is one of the oldest and most widely used additives in winemaking, with historical records dating back to Roman times. Its oenological functions can be summarized as follows [71, 64]:

1. *Antiseptic:* It inhibits the growth of microorganisms, showing greater activity against bacteria than yeasts. At low concentrations, the inhibition is reversible; at higher concentrations partial microbial death occurs.
2. *Antioxidant:* Slows oxidative reactions by reacting with dissolved oxygen, thereby preserving polyphenols and aromatic compounds.
3. *Antioxidase:* It deactivates or denatures oxidative enzymes (e.g., polyphenol oxidase) in the must, preventing pre-fermentative oxidation.

4. *Solubilizing agent:* At high concentrations, it favors the release of anthocyanins from grape skins by disrupting the vacuolar membrane.
5. *Binding agent:* It improves wine aroma and taste by reacting with volatile and pungent compounds (e.g., acetaldehyde, pyruvic acid), neutralizing their perception.
6. *Fining agent:* It facilitates the aggregation and removal of colloidal substances, contributing to clarification.

S. cerevisiae can naturally produce small amounts of SO_2 during alcoholic fermentation as an intermediate in the assimilatory reduction of sulfate to sulfite. However, these levels are generally insufficient to exert a significant protective effect [72], and rarely exceed 10 mg/L, though in some cases they may surpass 30 mg/L [64].

In practice, SO_2 is typically added in the form of potassium metabisulfite ($K_2S_2O_5$), which contains 57.6% active SO_2 by weight, Thus, 1 g of metabisulfite releases approximately 0.57 g of effective SO_2 [73]. Upon dissolution, SO_2 establishes chemical equilibria in aqueous solution, resulting in three main species: molecular SO_2 (the active form), bisulfite ion (HSO_3^-), and, to a lesser extent, sulfite ion (SO_3^{2-}). Their relative distribution is highly pH-dependent (Figure 2).

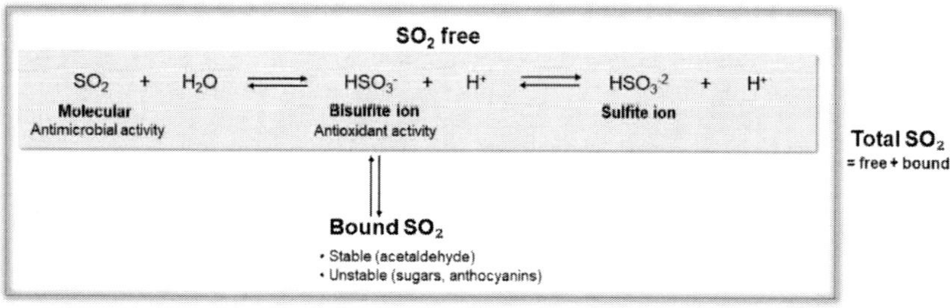

Figure 2. Dissociation of sulfur dioxide (SO_2) under different pH conditions.

At low pH, the equilibrium favors the presence of molecular SO_2, the primary antimicrobial agent. This form can freely diffuse through microbial cell membranes and disrupt vital cellular functions. Within the typical wine pH range (3-4), the bisulfite ion predominates and plays a key antioxidant role by reacting with oxidative intermediates (e.g., quinones, hydrogen peroxide), helping to preserve wine aroma, flavor, and color. The sulfite ion, although capable of reacting with oxygen, is present in negligible amounts at wine pH values and is highly unstable [64, 71, 74, 75].

Combined SO_2 species, those bound to sugars or carbonyl compounds such as aldehydes, are generally inactive in terms of both antimicrobial and antioxidant properties. Therefore, the concentration of free SO_2 (molecular SO_2 + HSO_3^-) is the most relevant parameter to monitor during winemaking. Moreover, lower pH levels increase the proportion of molecular SO_2, enhancing its preservative capacity.

To ensure consumer safety and product quality, legal limits for total SO_2 concentrations in wine have been established. According to the International Organisation of Vine and Wine (OIV), the maximum permissible total SO_2 concentration is 150 mg/L for dry red wines and 200 mg/L for dry white and rosé wines [74, 75]. Despite its technological benefits, SO_2 can have adverse health effects, particularly among sensitive individuals such as asthmatics, who may experience bronchospasms, respiratory difficulty, and coughing upon exposure. Other reported reactions include gastrointestinal symptoms, urticaria, angioedema, hypotension, shock, and, in rare cases, anaphylactic episodes [76].

In response, the World Health Organization (WHO) has set an acceptable daily intake (ADI) of 0.7 mg of SO_2 per kilogram of body weight (WHO, 2005) [77]. Consequently, regulatory frameworks require the labeling of wines containing more than 10 mg/L of total SO_2.

Growing consumer concern over sulfite safety and allergenicity has led to increased demand for alternative methods. In oenology, there is a clear trend toward SO_2 reduction, and recent viticultural research has increasingly focused on exploring alternative techniques and additives that can replicate their technological functions without compromising wine's quality [78].

3.2. Alternative Methods to SO_2 in Winemaking

The search for alternative methods to SO_2 responds to a combination of health concerns, consumer demands, labelling requirements, potential toxic effects, and a general trend towards the production of more "natural" wines with lower levels of additives. For this reason, the wine industry has promoted research into alternative methods that do not compromise the microbiological stability or quality of the wine. These strategies can be grouped into chemical, physical, and biological alternatives (Table 1).

Table 1. Alternatives methods proposed to replace SO_2 addition in wine-making

Methods		Mechanisms/Observations
Chemical	Dimethyl dicarbonate (DMDC)	Inhibits key enzymes in yeast glycolysis; rapidly hydrolyzed into methanol and CO_2; short-term effect.
	Lysozyme	Enzyme from egg white; effective against Gram-positive bacteria; inactive against yeasts and acetic acid bacteria; lacks antioxidant properties.
	Sorbic acid	Antifungal agent; prevents refermentation and biofilm-forming yeasts; ineffective against LAB, AAB, and spoilage yeasts; may produce off-flavors.
Physical	High Pressure Processing (HPP)	Applies 100-600 MPa to inactivate microbes; cold sterilization without thermal damage; effective but costly.
	Ultrasound	High-frequency waves cause cavitation; disrupts membranes; promotes phenolic extraction; may affect sensory properties.
	Ultraviolet irradiation	UV-C damages microbial DNA; effective on musts and bottled wines; limited by wine turbidity.
	Pulsed Electric Fields (PEF)	High-voltage pulses induce electroporation; reduces spoilage yeast; may require combination with other methods.
	Filtration	Physical retention of microbes using 0.45-0.2 μm membranes; reduces turbidity and spoilage risk.
Biological	Bacteriocins of LAB	Peptides produced by LAB; target Gram-positive bacteria; limited spectrum; often used with SO_2 or DMDC.
	Bioprotective yeasts	Selected *Saccharomyces* and non-*Saccharomyces* strains are used to outcompete spoilage microorganisms and enhance microbial stability. Antagonistic yeasts can produce antimicrobial peptides (AMPs) or killer toxins, which act as effective biocontrol agents that can be easily integrated into the winemaking process.

3.2.1. Chemical Alternatives

3.2.1.1. Dimethyl Dicarbonate (DMDC)

DMDC acts by inhibiting key enzymes involved in yeast glycolysis, although its effect on bacteria is limited [79]. In wine, DMDC hydrolyzes rapidly into methanol and carbon dioxide, producing an immediate but transient antimicrobial effect. Rapid homogenization after addition is critical for efficacy [78]. The methanol released remains within legal limits. DMDC use is authorized in the EU for wines with more than 5 g/L residual sugars.

3.2.1.2. Lysozyme

Lysozyme (muramidase or N-acetylmuramide glycanohydrolase) is an enzyme derived from egg white. It lyses the cell walls of Gram-positive bacteria [80] but is ineffective against Gram-negative bacteria and eukaryotic cells. Therefore, it does not control spoilage yeasts such as *Dekkera/Brettanomyces*, or acetic acid bacteria (Gram-negative). To date, no negative effects of lysozyme on the sensory effects in wine have been reported. Lysozyme is permitted up to 500 mg/L in must, and this dosage is cumulative in wine [81]. However, lysozyme cannot fully replace SO_2, as it lacks antioxidant properties and is inactive against yeasts and acetic bacteria.

3.2.1.3. Sorbic Acid

Sorbic acid, typically applied as potassium sorbate, is used to prevent refermentation by *S. cerevisiae* in wines with residual sugar and low-to-moderate alcohol. It also inhibits biofilm-forming yeasts such as *Candida* spp. However, it is ineffective against LAB, acetic acid bacteria, and certain spoilage yeasts [78], and is usually applied in combination with SO_2. In Europe, its use is authorized in salt form, with a maximum limit of 200 mg/L in the final product, due to potential health concerns.

3.2.2. Physical Alternatives

3.2.2.1. High Pressure Processing (HPP)

HPP involves subjecting wine or must very high pressures, ranging from 100 to 600 MPa, for a few minutes, which results in the destruction or inactivation of yeasts and bacteria. Microbial inactivation occurs due to rupture of cell membranes, protein denaturation, and loss of solutes [82]. Its main advantage lies in allowing cold sterilization, better preserving the aromatic and phenolic compounds of the wine compared to thermal methods.

This technique has shown good results in the inactivation of *Brettanomyces, Saccharomyces, Lactobacillus* and *O. oeni*, particularly when applied before bottling [78]. However, its high cost and the need for specialized equipment still limit its use.

3.2.2.2. Ultrasound

The use of high-frequency ultrasound (20-100 kHz) generates acoustic waves that induce cavitation in the liquid medium, causing the formation and collapse of microbubbles which produce high local pressures. The mechanism of microbial death is mainly due to thinning of cell membranes, localized heating, and the production of free radicals [83]. In winemaking, this technique can be used to reduce microbial load in the must prior to fermentation or to inactivate spoilage microorganisms. Its effectiveness depends on the power, duration, and temperature of the treatment; however, the sensory properties of the wine may be negatively affected by the appearance of oxidative, burnt, or smoky aromas.

3.2.2.3. Ultraviolet (UV) Irradiation

UV radiation includes wavelengths from the electromagnetic radiation spectrum ranging between 100 and 400 nm. UV-C rays (200-280 nm) are the most effective against microorganisms. UV-C light has strong germicidal activity, as it damages the DNA of microorganisms such as yeasts and bacteria, preventing their reproduction. Yeasts show greater resistance than bacteria, and among bacteria, Gram-negative are more sensitive. Its application in winemaking has been evaluated in musts to reduce the initial microbial load and allow better control of induced fermentations, or in bottled wines as a complementary treatment to ensure microbiological stability [84]. However, its effectiveness may be limited by wine turbidity, as UV radiation has low penetration.

3.2.2.4. Pulsed Electric Fields (PEF)

PEF technology consists of applying high-voltage electric pulses (5 to 50 kV/cm) for milliseconds on the liquid, generating pores in cell membranes (electroporation), thereby increasing permeability and leading to microbial death without significantly increasing temperature. In oenology, it has been studied for the inactivation of spoilage yeasts in wine before bottling, without negatively affecting the phenolic composition or color of the wine [85]. It is often combined with other techniques such as filtration or small doses of SO_2.

3.2.2.5. Filtration

Microfiltration is a well-established technique in the wine industry, especially during the final stages of the process. It consists of passing the wine through membranes with pores of 0.45 to 0.2 microns, which retain yeasts, bacteria, and suspended particles. Filtration can reduce microbial growth, browning, and turbidity through the removal of colloids [86].

3.2.3. Biological Alternatives

3.2.3.1. Lactic Acid Bacteria Producers of Bacteriocins

Bacteriocins are low molecular weight antimicrobial peptides, produced by certain strains of LAB, capable of inhibiting the growth of other bacterial strains, either of the same species or of different ones. Their mechanism of action involves disruption of the cytoplasmic membrane of sensitive bacteria by forming pores, which leads to the loss of ions and essential metabolites, dissipation of the proton-motive force, and ultimately, cell death [87]. These peptides are primarily active against Gram-positive bacteria and have minimal effect on Gram-negative species, yeasts, or molds. Therefore, their application is recommended in combination with SO_2 or other additives such as DMDC to achieve broader microbiological control [88]. Their strategic application can help regulate MLF and suppress spoilage LAB.

3.2.3.2. Yeasts with Bioprotective Effect

The use of selected yeasts, *Saccharomyces* and non-*Saccharomyces* species, represents an alternative biological strategy to reduce or replace the use of SO_2 in winemaking. These yeasts can exert a protective effect on the must or wine by limiting the development of undesirable microorganisms. Among the most studied non-*Saccharomyces* species, *M. pulcherrima* has shown the ability to inhibit several spoilage yeasts such as *B. bruxellensis* by production of pulcherrimin, a red pigment resulting from iron (Fe^{3+}) chelation [89].

On the other hand, various strains of *S. cerevisiae* have shown the ability to produce AMPs, distinct from killer toxins, which exhibit activity against spoilage yeasts and some bacteria [78]. Although these compounds are produced in low concentrations and have a more limited spectrum of action, the inoculation of selected starter cultures of *S. cerevisiae* is also important, as it reduces microbial competition and the risk of spoilage.

Several yeast strains, especially *S. cerevisiae* and some non-*Saccharomyces* species, can produce killer toxins, extracellular peptides or glycoproteins with specific antimicrobial activity against other sensitive yeast strains. They are particularly effective against yeasts of the different genera and can help control spoilage during alcoholic fermentation and wine storage. The use of killer strains could be considered a form sustainable of biocontrol that do not require modifications to the winemaking process and can be easily integrated with other oenological practices [14, 17, 26].

Regarding the use of killer strains as starter cultures in winemaking, Fernández de Ullivarri et al. [17] evaluated fermentative and biocontrol properties of two regional killer strains, *S. cerevisiae* Cf8 and *W. anomalus* Cf20, against potential spoilage yeasts under micro-vinifications conditions using Malbec must (Figure 3A) and white must (Figure 3B). The combination of both killer strains was effective controlling the growth of *Meyerozyma guilliermondii* Cd6 at a low SO_2 concentration (50 mg/L of metabisulfite) during fermentation.

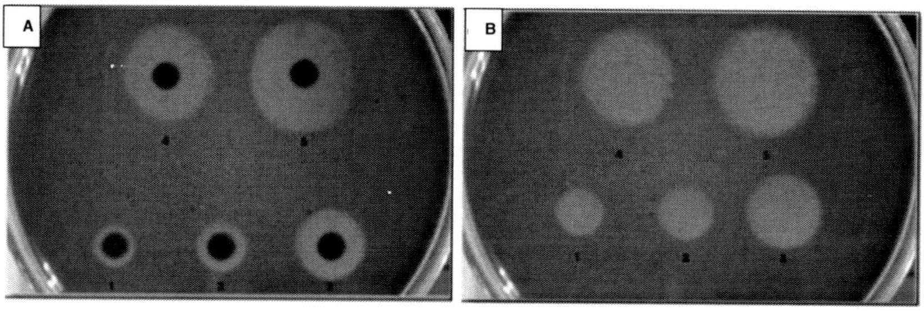

Figure 3. Killer activity of Malbec (A) and Torrontes (B) musts fermented by mixed culture of *S. cerevisiae/W. anomalus/M. guilliermondii* against *M. guilliermondii* Cd6 after 1, 2, 3, 4 and 5 days in YPD-MB pH 4.5 at 20°C. From Fernández de Ullivarri et al. 2024 [17].

Figure 4 shows the evolution of *M. guilliermondii* Cd6 populations in pure and mixed fermentations. Initially, the cell counts of Cd6 were similar in both the pure culture (Mg) and the fermentation with Cf8 (Sc+Mg). However, in the wine fermented by both killer strains (Sc+Wa+Mg), the cell population of *M. guilliermondii* Cd6 was reduced by 2 log units. At 3 days of fermentation, the strain Cd6 lost viability in both mixed trials, possibly due to a synergistic effect between killer toxins and ethanol. At the end of the process, a lower viability of Cd6 was observed in the wine Sc+Wa+Mg, which corresponded with higher killer activity (4×10^4 aU/mL) compared to the wine Sc+Mg (KA = 3.6×10^3 aU/mL). These findings suggest that the combination of killer non-*Saccharomyces* strains with *S. cerevisiae* as adjunct starters, could be a promising strategy for more effective control of wine spoilage microorganisms.

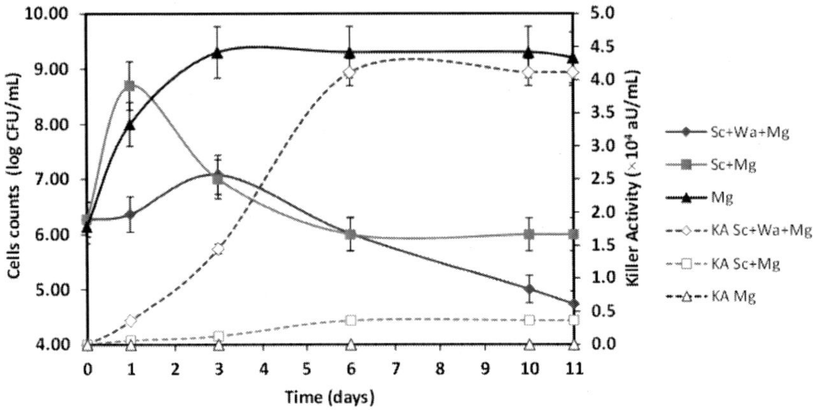

Figure 4. Viable cell counts of *M. guilliermondii* Cd6 (—) and killer activity (---) in Malbec wines obtained using mixed *S. cerevisiae/W. anomalus/M. guilliermondii* culture (◆,◇), mixed *S. cerevisiae/M. guilliermondii* culture (□,■) and pure *M. guilliermondii* culture (▲,△). Values represent the mean of two independent experiments. Linear vertical bars represent standard deviation. From Fernández de Ullivarri *et al*. 2024 [17].

4. Fruit Contamination: Use of Synthetic Fungicides and Yeasts to Control Deterioration

Fruits are susceptible to several spoilage agents particularly during post-harvest and storage stages. Among spoilage microorganisms, fungi are primarily responsible for the fruit deterioration, leading to economic losses and quality alterations. Postharvest losses due to fruit deterioration can account for 25–50% of total production, and shelf life is considerably shortened by microbial contamination [90, 91]. In addition, several fungal species belonging to *Alternaria*, *Aspergillus*, *Fusarium*, *Penicillium* genera can produce harmful secondary metabolites called mycotoxins (aflatoxins, OTA, patulin) in fruits [92]. In recent years, most studies have focused on the presence of *Aspergillus* and *Penicillium* species in wine, grapes, and grape juice contaminated with ochratoxin A [93, 94]. Patulin in apples and OTA in grapes are two mycotoxins currently at the center of research in fruits, whereas mycotoxins in citrus fruits are rarely detected. The use of antagonistic yeasts to remove or control these mycotoxins represents a potentially safe and environmentally friendly alternative [95].

One of the most important challenges in fruit preservation is the effective management of postharvest fungal pathogens. The use of synthetic fungicides to control postharvest diseases is a widespread and effective practice worldwide, and it is regulated by law [96]. Some fungicides, such as imazalil and thiobendazole, are highly effective in preventing postharvest diseases. Consequently, identifying a biofungicide with comparable efficacy remains a significant challenge. One approach could be combining microbial antagonists with reduced doses of fungicides, which represents a promising strategy for effective fungal biocontrol [90, 97].

Table 2. Antagonistic yeasts proposed as biocontrol agents to control fungal contamination in fruits

Antagonist	Fungi target	Fruit	Reference
Aerobasidium pullulans	*Aspergillus tubingensis*	Grape	[101]
	Botrytis cinerea	Apple	[102]
	Penicillium digitatum	Mandarin	[103]
	P. expansum	Pear, Grape	[104]
Candida membranifaciens	*P. digitatum*	Orange	[105]
Clavispora lusitaniae	*P. digitatum*	Lemon	[13]
Cryptococcus laurentii	*B. cinerea*	Strawberry	[106]
Hanseniaspora uvarum	*B. cinerea*	Grape	[107]
Kluyveromyces marxianus	*P. expansum*	Apple	[108]
Metschnikowia pulcherrima	*Alternaria alternata*	Grape	[109]
	B. cinerea		[103]
Pichia fermentans	*P. digitatum*	Lemon	[13]
Saccharomyces cerevisiae	*B. cinerea*	Grape	[103]
	P. digitatum	Lemon	[13]
Starmerella bacillaris	*B. cinerea*	Grape	[110]
Wickerhamomyces anomalus	*B. cinerea*	Grape	[103]
	P. italicum	Strawberry	

Among the antagonistic microorganisms used in fruit biocontrol, yeasts offer some advantages over bacteria and filamentous fungi in preventing postharvest diseases [98]. In general, yeasts are more resistant to environmental stress, can develop in fruits with high sugar or acid content, grow on simple substrates, and do not produce toxic secondary metabolites or allergenic spores [99, 100]. Although yeasts that naturally occur on fruit surfaces are most commonly used for the biocontrol of postharvest diseases, antagonistic yeasts isolated from other ecosystems were also studied [99]. The most extensively studied yeasts with antagonistic activity belong to the genera *Aureobasidium*, *Candida*, *Metschnikowia*, *Pichia*, *Saccharomyces* and *Wickerhamomyces* (Table 2).

Several antagonistic yeasts have achieved commercial success in the agricultural market. There are currently yeast-based biofungicides such as Aspire™ (*Candida oleophila*), BoniProtect™ (*Aureobasidium pullulans*), Noli™ (*Metschnikowia fructicola*), and Romeo™ (*S. cerevisiae*) that were registered for use in Europe and the United States [90, 91]. In Argentina, the commercialization of antagonistic yeasts remains an unresolved issue. Although numerous studies have demonstrated the effectiveness of yeasts as biofungicides for controlling fungal diseases in table grapes, wine grapes, lemons, and other fruits, the next steps toward market development are still pending [13, 104, 109].

5. Antagonistic Yeasts as Biocontrol Agents: Mechanisms and Applications

Chemical preservatives have long served as primary tools for microbial control in the food and beverage industry. Their principal role is to inhibit spoilage organisms and foodborne pathogens, extending product shelf life and reducing the risk of contamination-related losses [111]. Despite their utility, chemical preservatives also pose multiple challenges such as emergence of resistance, sensory impact, interactions with food components and consumer perception considering the growing demand for "clean-label" products and environmental concerns [112, 113].

In this context, biological control using antagonistic yeasts is a promising strategy for managing spoilage microorganisms in food and beverage systems. These yeasts exhibit natural inhibitory activity against bacteria, spoilage fungi, and competing yeasts through diverse mechanisms, offering an eco-friendly alternative to chemical preservatives [15]. Antagonistic yeasts have been studied and applied in various stages of the food production chain, including fermentation, storage, and post-harvest management, with documented effectiveness in wine, fruit, and minimally processed products [13, 17, 24, 25]. The interest in antagonistic yeasts is driven by increasing consumer demand for clean-label and additive-free products, as well as regulatory limitations on traditional preservatives like sulfur dioxide, sorbates, and benzoates [113]. Research in this field has accelerated due to advances in microbial ecology, molecular biology, and omics-based techniques, allowing the detailed characterization of antagonistic mechanisms and strain-specific activities [114].

5.1. Mechanisms of Antagonism

Antagonistic yeasts employ a diverse set of antimicrobial strategies, grouped into five major categories:

5.1.1. Killer Toxins (Mycocins)

One of the most studied mechanisms is the production of killer toxins, proteinaceous or glycoprotein molecules, secreted by certain yeast strains that kill other sensitive yeast or fungal species. These toxins typically act by disrupting membrane integrity, inhibiting cell wall synthesis, or blocking essential enzymatic pathways [26]. They function optimally in acidic environments, compatible with wine, fruit juice, and other fermented foods [115].

Killer toxin genes may reside on chromosomes, plasmids, or double-stranded RNA virus-like elements, and producing yeasts are immune to their own toxins via regulated immunity pathways [26].

Notable examples include:

- *W. anomalus* (formerly *Pichia anomala*), producing killer toxins active against *Brettanomyces*, *Pichia, Meyerozyma*, and *Aspergillus* species [12, 17].
- Killer strains of *S. cerevisiae* (e.g., K1, K2, K28 types) used to suppress spoilage yeasts in winemaking [116].
- *Pichia membranifaciens*, which secretes heat-stable killer toxins effective against *Debaryomyces hansenii* and other post-harvest spoilage yeasts [117].

5.1.2. Nutrient Competition and Space Exclusion

Yeasts can inhibit microbial competitors by rapidly colonizing ecological niches and consuming essential nutrients. This mechanism, though indirect, is powerful in food systems with limited resources [118]. *M. pulcherrima* is particularly effective due to its production of pulcherrimin, an iron-chelating pigment that renders iron unavailable to competing microbes [89]. Other yeasts outcompete spoilage organisms by quickly utilizing available sugars, nitrogen, or vitamins, particularly during early fermentation or storage. Colonization of fruit wounds and storage surfaces by antagonistic yeasts also prevents the establishment of pathogens like *Penicillium expansum* and *B. cinerea* [27].

5.1.3. Volatile Organic Compounds (VOCs)

Some antagonistic yeasts release VOCs with antimicrobial effects. These include:

- 2-phenylethanol
- Ethyl acetate
- Acetaldehyde
- Isoamyl alcohol

These compounds interfere with membrane function, disrupt metabolic activity, or inhibit germination and sporulation in molds. For example, *M. pulcherrima* and *H. uvarum* produce VOCs that inhibit *B. cinerea* and *A. niger* during post-harvest storage of grapes and apples [11]. In sealed or low-ventilation environments, VOCs can accumulate and exert strong antifungal pressure [119].

5.1.4. Secretion of Hydrolytic Enzymes

Yeasts such as *Candida oleophila*, *P. membranifaciens*, and *Cryptococcus laurentii* produce extracellular lytic enzymes like chitinases, β-1,3-glucanases, and proteases [120]. These enzymes degrade fungal cell walls or interfere with pathogen colonization.

Enzymatic activity often acts synergistically with other mechanisms, enhancing yeast biocontrol efficacy under stress conditions or in damaged fruit tissue [121].

5.1.5. Organic Acid Production and pH Modification

Some yeasts lower environmental pH through organic acid production (e.g., acetic, lactic), which suppresses the growth of pH-sensitive microbes. This is particularly relevant in fermented vegetables, soft drinks, and fruit-based beverages. In mixed fermentations, acidification by non-*Saccharomyces* yeasts modulates microbial succession and enhances spoilage control without external acidulants [122].

A wide range of antagonistic yeasts have been investigated and applied in food systems. Many of these yeasts are native to food environments and have been granted GRAS or QPS status by regulatory agencies, facilitating their use in commercial applications (Table 3).

Table 3. Relevant yeast species as biocontrol agents

Species	Notable Activity	Application
W. anomalus	Killer toxin, VOCs	Wine, grapes, berries
M. pulcherrima	Iron depletion, VOCs	Grapes, apples, berries
P. membranifaciens	Killer toxin, enzymes	Citrus, apples
C. oleophila	Surface colonization	Apples, citrus (Aspire)
Aureobasidium pullulans	Protease, competition	Grapes, blueberries (BoniProtect)
H. uvarum	VOCs	Grapes, cider
Torulaspora delbrueckii	Killer toxin, fermentation compatibility	Wine, cider

5.2. Applications in Food Systems

Figure 5 shows some applications of antagonistic yeasts in the production of fermented beverages and the postharvest preservation of fruits.

5.2.1. Wine-making and Fermented Beverages

In wine, biocontrol yeasts can reduce *Brettanomyces* proliferation, volatile phenol formation, and acetic acid production. *M. pulcherrima* and *T. delbrueckii* are used in sequential fermentations to dominate early stages, followed by inoculation with *S. cerevisiae* to complete fermentation [123, 124]. Killer *S. cerevisiae* strains also suppress spoilage yeasts in spontaneous fermentations while preserving desired aroma profiles [17, 125].

5.2.2. Post-Harvest Fruit Preservation

Biocontrol yeasts are applied post-harvest via sprays, dips, or wax coatings to protect fruits from decay. *C. oleophila*, *A. pullulans*, and *P. fermentans* have demonstrated consistent protection against *P. expansum*, *B. cinerea*, and *Alternaria alternata* on apples, pears, grapes, and citrus fruits [126-127]. Their activity persists during cold storage and transportation, making them suitable for integration into supply chains.

Advantages

- *Selective action:* Antagonists target specific spoilage organisms while preserving fermentative flora.
- *Low toxicity:* No allergenic residues or harmful metabolites.
- *Stress tolerance:* Survive and act in low pH, low temperature, high osmolarity.

- *Compatibility:* Can be integrated with physical or mild chemical hurdles.
- *Regulatory approval:* Many species have GRAS or QPS status.
- *Alignment with consumer trends:* Meet clean-label and natural preservation demands.

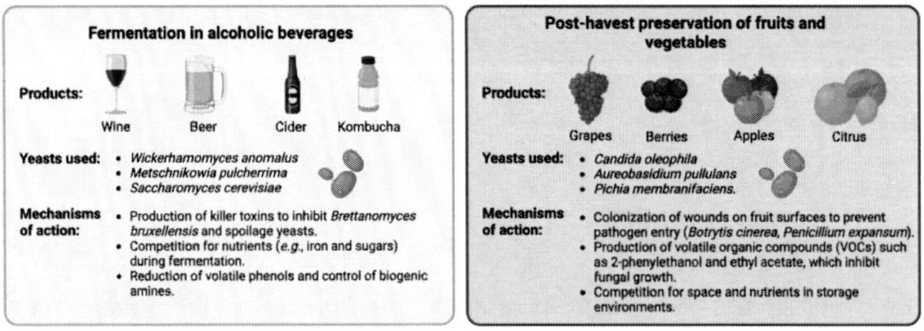

Figure 5. Applications of yeasts as biocontrol agents in food and beverage industry.

Limitations

- *Strain specificity:* Antimicrobial activity varies significantly between strains of the same species.
- *Formulation challenges:* Maintaining cell viability in dry or commercial formulations requires technological investment.
- *Regulatory barriers:* Approvals for new strains may require safety testing.
- *Inconsistent efficacy:* Lab results may not always translate to real food matrices due to native microbiota and environmental variability.
- *Shelf-life and stability:* Decreased viability over time without proper encapsulation or storage.

Future Perspectives

- *Genomic and transcriptomic profiling* to identify new antimicrobial pathways and optimize strain selection.
- *Microbial consortia:* Use of synergistic yeast-bacteria combinations to broaden protection.
- *Encapsulation technologies:* Spray-drying, freeze-drying, and lipid carriers for improved delivery and stability.

- *Smart application systems:* Incorporating antagonistic yeasts into biofilms, coatings, or fermentative starter cultures.
- *Regulatory harmonization:* Streamlining approval processes to enable global use of safe, well-characterized strains.

The continued integration of yeast-based biocontrol into food safety systems will depend on a multidisciplinary approach combining microbiology, food technology, and industrial-scale formulation development.

Conclusion

Biological control using antagonistic yeasts represents a viable and scientifically supported alternative to synthetic preservatives in modern food systems. These yeasts exhibit diverse and complementary mechanisms, including killer toxin production, competition for nutrients, secretion of volatile and hydrolytic compounds, and pH modulation, that enable them to inhibit a wide range of spoilage organisms across various food matrices.

Compared to chemical preservatives, antagonistic yeasts offer greater specificity, minimal sensory interference, lower toxicological risk, and greater compatibility with clean-label trends. They are particularly promising in fermentation (e.g., wine, kombucha), post-harvest fruit preservation, and minimally processed foods. Some species, such as *W. anomalus*, *M. pulcherrima*, and *C. oleophila*, have already been adopted in commercial applications.

Despite their advantages, challenges persist. These include variability in strain efficacy, limited regulatory harmonization, difficulty in formulation stability, and inconsistent results in complex food matrices. Nevertheless, recent advances in microbial genomics, metabolomics, and encapsulation technologies are rapidly addressing these gaps.

In conclusion, antagonistic yeasts are poised to play a central role in the future of sustainable food preservation. Their integration with physical and mild chemical hurdles will likely enable the design of multifunctional, low-impact biocontrol systems tailored to specific spoilage risks and consumer expectations.

References

[1] Pitt, J. I., & Hocking, A. D. (2022). Ecology of fungal food spoilage. *In Fungi and food spoilage* (pp. 3-12). Cham: Springer International Publishing.

[2] Snyder, A. B., Martin, N., & Wiedmann, M. (2024). Microbial food spoilage: Impact, causative agents and control strategies. *Nature Reviews Microbiology, 22(9)*, 528-542.

[3] Pinu, F. R. (2016). Early detection of food pathogens and food spoilage microorganisms: Application of metabolomics. *Trends in Food Science & Technolo*gy, 54, 213-215.

[4] Campêlo, M. C. S., Medeiros, J. M. S., & Silva, J. B. A. (2019). Natural products in food preservation. *International Food Research Journal*, *26*(1).

[5] Batiha, G. E. S., Hussein, D. E., Algammal, A. M., et al. (2021). Application of natural antimicrobials in food preservation: Recent views. *Food Control*, *126*, 108066.

[6] Taylor, T. M., Ravishankar, S., Bhargava, K., & Juneja, V. K. (2019). Chemical preservatives and natural food antimicrobials. *Food microbiology: Fundamentals and frontiers*, 705-731.

[7] Mashabela, M. N., Ndhlovu, P. T., & Mbeng, W. O. (2022). Herbs and spices' antimicrobial properties and possible use in the food sector. In *Herbs and Spices-New Advances*. IntechOpen.

[8] Pinto, L., Tapia-Rodríguez, M. R., Baruzzi, F., & Ayala-Zavala, J. F. (2023). Plant antimicrobials for food quality and safety: Recent views and future challenges. *Foods*, *12*(12), 2315.

[9] Shi, C., & Maktabdar, M. (2022). Lactic acid bacteria as biopreservation against spoilage molds in dairy products- A review. *Frontiers in microbiology*, *12*, 819684.

[10] Agarbati, A., Canonico, L., Pecci, T., Romanazzi, G., Ciani, M., & Comitini, F. (2022). Biocontrol of non-*Saccharomyces* yeasts in vineyard against the gray mold disease agent *B. cinerea. Microorganisms*, *10*(2), 200.

[11] Oztekin, S., & Karbancioglu-Guler, F. (2024). Recruiting grape-isolated antagonistic yeasts for the sustainable bio-management of *B. cinerea* on grapes. *Food and Energy Security*, *13*(1), e528.

[12] Fernandez de Ullivarri, M., Mendoza, L. M., & Raya, R. R. (2018). Characterization of the killer toxin KTCf20 from *W. anomalus*, a potential biocontrol agent against wine spoilage yeasts. *Biological control*, *121*, 223-228.

[13] Pereyra, M. M., Díaz, M. A., Vero, S., & Dib, J. R. (2024). Enhancing biological control of post-harvest green mold in lemons: Synergistic efficacy of native yeasts with diverse mechanisms of action. *Plos one*, *19*(4), e0301584.

[14] Vepštaitė-Monstavičė, I., Lukša-Žebelovič, J., Apšegaitė, V., Mozūraitis, R., Lisicinas, R., Stanevičienė, R., ... & Servienė, E. (2025). Profiles of killer systems and volatile organic compounds of rowanberry and rosehip-inhabiting yeasts substantiate implications for biocontrol. *Foods*, *14*(2), 288.

[15] Sipiczki, M. (2023). Identification of antagonistic yeasts as potential biocontrol agents: Diverse criteria and strategies. *International Journal of Food Microbiology*, *406*, 110360.

[16] Janisiewicz, W. (2020). Biological control of diseases of fruits. In *Biocontrol of plant diseases* (pp. 153-165). CRC Press.

[17] Fernandez de Ullivarri, M., Merín, M. G., Raya, R. R., de Ambrosini, V. I. M., & Mendoza, L. M. (2024). Killer yeasts used as starter cultures to modulate the behavior of potential spoilage non-*Saccharomyces* yeasts during Malbec wine fermentation. *Food Bioscience*, *57*, 103424.

[18] Çomitini, F., Canonico, L., Agarbati, A., & Ciani, M. (2023). Biocontrol and probiotic function of non-*Saccharomyces* yeasts: New insights in agri-food industry. *Microorganisms*, *11*(6), 1450.

[19] El Dana, F., David, V., Hallal, M. A., Tourdot-Maréchal, R., Hayar, S., Colosio, M. C., & Alexandre, H. (2025). *M. pulcherrima* and *Lachancea thermotolerans* Killer Toxins: Contribution to Must Bioprotection. *Foods*, *14*(9), 1462.

[20] Liu, Z., Du, S., Ren, Y., & Liu, Y. (2018). Biocontrol ability of killer yeasts (*S. cerevisiae*) isolated from wine against *Colletotrichum gloeosporioides* on grape. *Journal of Basic Microbiology*, *58*(1), 60-67.

[21] Perez, M. F., Isas, A. S., Aladdin, A., El Enshasy, H. A., & Dib, J. R. (2018). Killer yeasts as biocontrol agents of post-harvest fungal diseases in lemons. *Sustainable Technologies for the Management of Agricultural Wastes*, 87-98.

[22] Mannazzu, I., Domizio, P., Carboni, et al. (2019). Yeast killer toxins: From ecological significance to application. *Critical reviews in biotechnology*, *39*(5), 603-617.

[23] Singh, N., & Gaur, S. (2021). GRAS fungi: a new horizon in safer food product. *Fungi in sustainable food production*, 27-37.

[24] Villalba, M. L., Lutz, M. C., Lopez, S., Pildain, M. B., & Sangorrín, M. P. (2016). Patagonian antagonist yeasts for food biopreservation. *Biology and Biotechnology of Patagonian Microorganisms*, 301-323.

[25] de Melo Pereira, G. V., de Mello Sampaio, V., Wiele, N., et al. (2024). How yeast has transformed the coffee market by creating new flavors and aromas through modern post-harvest fermentation systems. *Trends in Food Science & Technology*, 104641.

[26] Klassen, R., Schaffrath, R., Buzzini, P., & Ganter, P. F. (2017). Antagonistic interactions and killer yeasts. *Yeasts in natural ecosystems: Ecology*, 229-275.

[27] Freimoser, F. M., Rueda-Mejia, M. P., Tilocca, B., & Migheli, Q. (2019). Biocontrol yeasts: mechanisms and applications. *World Journal of Microbiology and Biotechnology*, *35*(10), 154.

[28] Jagadeesan, B., Gerner-Smidt, P., Allard, M. W., et al. (2019). The use of next generation sequencing for improving food safety: Translation into practice. *Food microbiology*, *79*, 96-115.

[29] Stojanović, Z., & Kos, J. (2020). Detection of metabolites of microbial origin in beverages with harmful effect on human health—Biogenic amines and mycotoxins. In *Safety issues in beverage production* (pp. 39-77). Academic Press.

[30] Capozzi, V., Fragasso, M., Romaniello, R., Berbegal, C., Russo, P., & Spano, G. (2017). Spontaneous food fermentations and potential risks for human health. *Fermentation*, *3*(4), 49.

[31] Cosme, F., Vilela, A., Filipe-Ribeiro, L., Inês, A., & Nunes, F. M. (2018). Wine microbial spoilage: advances in defects remediation. In *Microbial contamination and food degradation* (pp. 271-314). Academic press.

[32] Barboráková, Z., Jakabová, S., Maskova, Z., Mrvová, M., Uzsáková, V., Maková, J., & Tancinova, D. (2023). Toxin producing micromycetes of the genus *Penicillium* and *Aspergillus* on berries, grapes, and tomato fruits in Slovak stores. *Journal of Microbiology, Biotechnology and Food Sciences*, *13*(1), e9927-e9927.

[33] Lorenzini, M., Cappello, M. S., Perrone, G., Logrieco, A., & Zapparoli, G. (2019). New records of *Penicillium* and *Aspergillus* from withered grapes in Italy, and description of *Penicillium fructuariae-cellae* sp. nov.*Phytopathologia Mediterranea*, *58*(2), 323-340.

[34] Dean R, Van Kan JA, Pretorius ZA, et al. The top 10 fungal pathogens in molecular plant pathology. *Mol Plant Pathol.* 2012;13(4):414–430.

[35] Dwivedi, M., Singh, P., & Pandey, A. K. (2024). Botrytis fruit rot management: What have we achieved so far? *Food Microbiology*, 104564.

[36] Tjamos, S. E., Antoniou, P. P., Kazantzidou, A., Antonopoulos, D. F., Papageorgiou, I., & Tjamos, E. C. (2004). *Aspergillus niger* and *Aspergillus carbonarius* in Corinth raisin and wine-producing vineyards in Greece: population composition, ochratoxin A production and chemical control. *Journal of Phytopathology*, *152*(4), 250-255.

[37] Suárez, R., Suárez-Lepe, J. A., Morata, A., & Calderón, F. (2007). The production of ethylphenols in wine by yeasts of the genera *Brettanomyces* and *Dekkera*: A review. *Food Chemistry*, *102*(1), 10-21.

[38] Lleixà, J., Martínez-Safont, M., Masneuf-Pomarede, I., Magani, M., Albertin, W., Mas, A., & Portillo, M. C. (2021). Genetic and phenotypic diversity of *B. bruxellensis* isolates from ageing wines. *Food Bioscience*, *40*, 100900.

[39] Costantini, A., Cersosimo, M., Del Prete, V., & Garcia-Moruno, E. (2006). Production of biogenic amines by lactic acid bacteria: screening by PCR, thin-layer chromatography, and high-performance liquid chromatography of strains isolated from wine and must. *Journal of Food Protection*, *69*(2), 391-396.

[40] Turna, N. S., Chung, R., & McIntyre, L. (2024). A review of biogenic amines in fermented foods: Occurrence and health effects. *Heliyon*, *10*(2).

[41] Costantini, A., Pietroniro, R., Doria, F., Pessione, E., & Garcia-Moruno, E. (2013). Putrescine production from different amino acid precursors by lactic acid bacteria from wine and cider. *International Journal of Food Microbiology*, *165*(1), 11-17.

[42] Montanari, C., Tabanelli, G., Zamagna, I., et al. (2019). Modeling of yeast thermal resistance and optimization of the pasteurization treatment applied to soft drinks. *International Journal of Food Microbiology*, *301*, 1-8.

[43] Rawsthorne, H., & Phister, T. G. (2006). A real-time PCR assay for the enumeration and detection of *Z. bailii* from wine and fruit juices. *International Journal of Food Microbiology*, *112*(1), 1-7.

[44] Stratford, M., Steels, H., Nebe-von-Caron, G., Novodvorska, et al. (2013). Extreme resistance to weak-acid preservatives in the spoilage yeast *Z. bailii*. *International Journal of Food Microbiology*, *166*(1), 126-134.

[45] Alonso, V. P. P., Furtado, M. M., Iwase, C. H. T., Brondi-Mendes, J. Z., & Nascimento, M. D. S. (2024). Microbial resistance to sanitizers in the food industry. *Critical Reviews in Food Science and Nutrition*, *64*(3), 654-669.

[46] Nkemngong, C., & Teska, P. (2024). Biofilms, mobile genetic elements and the persistence of pathogens on environmental surfaces in healthcare and food processing environments. *Frontiers in Microbiology*, *15*, 1405428.

[47] Odeyemi, O. A., Alegbeleye, O. O., Strateva, M., et al. (2020). Understanding spoilage microbial community and spoilage mechanisms in foods of animal origin. *Comprehensive Reviews in Food Science and Food Safety*,*19*(2), 311-331.

[48] Fernane, F., Cano-Sancho, G., Sanchis, V., Marín, S., & Ramos, A. J. (2010). Aflatoxins and ochratoxin A in pistachios sampled in Spain: Occurrence and presence of mycotoxigenic fungi. *Food Additives and Contaminants*, *3*(3), 185-192.

[49] EFSA Panel on Contaminants in the Food Chain (CONTAM), Schrenk, D., Bodin, L., et al. (2020). Risk assessment of ochratoxin A in food. *EFSA journal*, *18*(5), e06113.

[50] Abi Assaf, J., Holden, E. R., Trampari, E., et al. (2024). Common food preservatives impose distinct selective pressures on *Salmonella Typhimurium* planktonic and biofilm populations. *Food Microbiology*, *121*, 104517.

[51] Smith, J. R., Martini, A., & Gibney, P. A. (2024). Utilizing functional genomics in *S. cerevisiae* to characterize food preservative compounds: A pilot study. *Journal of Food Science*, *89*(2), 1196-1210.

[52] Mafe, A. N., Edo, G. I., Makia, R. S., et al. (2024). A review on food spoilage mechanisms, food borne diseases and commercial aspects of food preservation and processing. *Food Chemistry Advances*, *5*, 100852.

[53] De Filippis, F., La Storia, A., Villani, F., & Ercolini, D. (2019). Strain-level diversity analysis of *Pseudomonas fragi* after in situ pangenome reconstruction shows distinctive spoilage-associated metabolic traits clearly selected by different storage conditions. *Applied and environmental microbiology*, *85*(1), e02212-18.

[54] Pretorius, I. S. (2000). Tailoring wine yeast for the new millennium: novel approaches to the ancient art of winemaking. *Yeast*, *16*(8), 675-729.

[55] Ribéreau-Gayon, P., & Traduction, A. (2000). *Vol. 2: The chemistry of wine: stabilization and treatments*. Chichester [etc.]: Wiley.

[56] Fleet, G. H. (2007). Yeasts in foods and beverages: impact on product quality and safety. *Current opinion in biotechnology*, *18*(2), 170-175.

[57] Bartowsky, E. J., & Borneman, A. R. (2011). Genomic variations of *Oenococcus oeni* strains and the potential to impact on malolactic fermentation and aroma compounds in wine. *Applied Microbiology and Biotechnology*, *92*, 441-447.

[58] Bartowsky, E. J., Costello, P. J., & Chambers, P. J. (2015). Emerging trends in the application of malolactic fermentation. *Australian Journal of Grape and Wine Research*, *21*, 663-669.

[59] Virdis, C., Sumby, K., Bartowsky, E., & Jiranek, V. (2021). Lactic acid bacteria in wine: Technological advances and evaluation of their functional role. *Frontiers in Microbiology*, *11*, 612118.

[60] Mortimer, R. K. (2000). Evolution and variation of the yeast (*Saccharomyces*) genome. *Genome Research*, *10*(4), 403-409.

[61] Santamaría, P., Garijo, P., López, R., Tenorio, C., & Gutiérrez, A. R. (2005). Analysis of yeast population during spontaneous alcoholic fermentation: effect of the age of the cellar and the practice of inoculation. *International Journal of Food Microbiology*, *103*(1), 49-56.

[62] Knight, S., Klaere, S., Fedrizzi, B., & Goddard, M. R. (2015). Regional microbial signatures positively correlate with differential wine phenotypes: evidence for a microbial aspect to terroir. *Scientific Reports*, *5*(1), 14233.

[63] Gonzalez, R., & Morales, P. (2022). Truth in wine yeast. *Microbial Biotechnology*, *15*(5), 1339-1356.

[64] Ribéreau-Gayon, P., Dubourdieu, D., Donèche, B., & Lonvaud, A. (Eds.). (2006). *Handbook of enology, Volume 1: The microbiology of wine and vinifications* (Vol. 1). John Wiley & Sons.

[65] Smyth, M., Russell, J., Milanowski, T., & Milanowski, T. (2011). Introduction to Winemaking. *Solar Energy in the Winemaking Industry*, 19-46.

[66] Agnolucci, M., Tirelli, A., Cocolin, L., & Toffanin, A. (2017). *Brettanomyces bruxellensis* yeasts: Impact on wine and winemaking. *World Journal of Microbiology and Biotechnology*, *33*, 1-6.

[67] Caruso, M., Fiore, C., Contursi, M., Salzano, G., Paparella, A., & Romano, P. (2002). Formation of biogenic amines as criteria for the selection of wine yeasts. *World Journal of Microbiology and Biotechnology*, *18*, 159-163.

[68] Du Toit, W. D., Pretorius, I. S., & Lonvaud-Funel, A. (2005). The effect of sulphur dioxide and oxygen on the viability and culturability of a strain of *Acetobacter pasteurianus* and a strain of *Brettanomyces bruxellensis* isolated from wine. *Journal of applied microbiology*, *98*(4), 862-871.

[69] Ortiz-Villeda, B., Lobos, O., Aguilar-Zuniga, K., & Carrasco-Sánchez, V. (2021). Ochratoxins in wines: A review of their occurrence in the last decade, toxicity, and exposure risk in humans. *Toxins*, *13*(7), 478.

[70] Sumby, K. M., Grbin, P. R., & Jiranek, V. (2014). Implications of new research and technologies for malolactic fermentation in wine. *Applied Microbiology and Biotechnology*, *98*, 8111-8132.

[71] Giacosa, S., Segade, S. R., Cagnasso, E., Caudana, A., Rolle, L., & Gerbi, V. (2019). SO_2 in wines: Rational use and possible alternatives. In *Red wine technology* (pp. 309-321). Academic Press.

[72] Thomas, D., & Surdin-Kerjan, Y. (1997). Metabolism of sulfur amino acids in *Saccharomyces cerevisiae*. *Microbiology and Molecular Biology Reviews*, *61*(4), 503-532.

[73] Henderson, P. (2009). Sulfur dioxide: science behind this antimicrobial, anti-oxidant wine additive. *Practical Winery & Vineyard Journal*, *1*, 1-7.

[74] Boulton, R. B., Singleton, V. L., Bisson, L. F., & Kunkee, R. E. (2013). *Principles and practices of winemaking*. Springer Science & Business Media.

[75] Waterhouse, A. L., Sacks, G. L., & Jeffery, D. W. (2024). *Understanding wine chemistry*. John Wiley & Sons.

[76] Vally, H., Misso, N. L., & Madan, V. (2009). Clinical effects of sulphite additives. *Clinical & Experimental Allergy*, *39*(11), 1643-1651.

[77] WHO (World Health Organization). (2005). Air quality guidelines for particulate matter, ozone, nitrogen dioxide and sulfur dioxide.

[78] Lisanti, M. T., Blaiotta, G., Nioi, C., & Moio, L. (2019). Alternative methods to SO_2 for microbiological stabilization of wine. *Comprehensive Reviews in Food Science and Food Safety*, *18*(2), 455-479.

[79] Costa, A., Barata, A., Malfeito-Ferreira, M., & Loureiro, V. (2008). Evaluation of the inhibitory effect of dimethyl dicarbonate (DMDC) against wine microorganisms. *Food microbiology*, *25*(2), 422-427.

[80] Liburdi, K., Benucci, I., & Esti, M. (2014). Lysozyme in wine: an overview of current and future applications. *Comprehensive Reviews in Food Science and Food Safety*, *13*(5), 1062-1073.

[81] Chen, K., Han, S. Y., Zhang, B., Li, M., & Sheng, W. J. (2015). Development of lysozyme-combined antibacterial system to reduce sulfur dioxide and to stabilize I talian R iesling ice wine during aging process. *Food science & nutrition*, *3*(5), 453-465.

[82] Mok, C., Song, K. T., Park, Y. S., Lim, S., Ruan, R., & Chen, P. (2006). High hydrostatic pressure pasteurization of red wine. *Journal of food science*, *71*(8), M265-M269.

[83] Chemat, F., & Khan, M. K. (2011). Applications of ultrasound in food technology: Processing, preservation and extraction. *Ultrasonics sonochemistry*, *18*(4), 813-835.

[84] Falguera, V., Forns, M., & Ibarz, A. (2013). UV–vis irradiation: An alternative to reduce SO2 in white wines?. *LWT-Food Science and Technology*, *51*(1), 59-64.

[85] González-Arenzana, L., Portu, J., López, R., López, N., Santamaría, P., Garde-Cerdán, T., & López-Alfaro, I. (2015). Inactivation of wine-associated microbiota by continuous pulsed electric field treatments. *Innovative Food Science & Emerging Technologies*, *29*, 187-192.

[86] Silva, F. V., & Van Wyk, S. (2021). Emerging non-thermal technologies as alternative to SO2 for the production of wine. *Foods*, *10*(9), 2175.

[87] Jack, R. W., Tagg, J. R., & Ray, B. (1995). Bacteriocins of gram-positive bacteria. *Microbiological reviews*, *59*(2), 171-200.

[88] Rojo-Bezares, B., Sáenz, Y., Zarazaga, M., Torres, C., & Ruiz-Larrea, F. (2007). Antimicrobial activity of nisin against *Oenococcus oeni* and other wine bacteria. *International journal of food microbiology*, *116*(1), 32-36.

[89] Puyo, M., Simonin, S., Bach, B., Klein, G., Alexandre, H., & Tourdot-Maréchal, R. (2023). Bio-protection in oenology by *Metschnikowia pulcherrima*: from field results to scientific inquiry. *Frontiers in Microbiology*, *14*, 1252973.

[90] Agirman, B., Carsanba, E., Settanni, L., & Erten, H. (2023) Exploring yeast-based microbial interactions: The next frontier in postharvest biocontrol. *Yeast*, 40(10), 457-475.

[91] Oztekin, S., Dikmetas, D. N., Devecioglu, D., Acar, E. G., & Karbancioglu-Guler, F. (2023) Recent Insights into the Use of Antagonistic Yeasts for Sustainable Biomanagement of Postharvest Pathogenic and Mycotoxigenic Fungi in Fruits with

Their Prevention Strategies against Mycotoxins. *Journal of Agriculture and Food Chem*istry, 71(26), 9923-9950.

[92] Sanzani, S. M., Reverberi, M., & Geisen, R. (2016) Mycotoxins in Harvested Fruits and Vegetables: Insights in Producing Fungi, Biological Role, Conducive Conditions, and Tools to Manage Postharvest Contamination. *Postharvest Biol. Technol.*, *122*, 95-105.

[93] Giovannoli, C., Passini, C., Di Nardo, F., Anfossi, L., & Baggiani, C. (2014) Determination of Ochratoxin A in Italian Red Wines by Molecularly Imprinted Solid Phase Extraction and HPLC Analysis. *J. Agric. Food Chem.*, *62 (22)*, 5220-5225.

[94] Mandappa, I. M., Basavaraj, K., & Manonmani, H. K. (2018) Analysis of Mycotoxins in Fruit Juices; Elsevier Inc.: Amsterdam.

[95] Sharma, V., & Patial, V. (2021) Food Mycotoxins: Dietary Interventions Implicated in the Prevention of Mycotoxicosis. ACS *Food Sci. Technol.*, *1 (10)*, 1717−1739.

[96] de Paiva, E., Serradilla, M. J., Ruiz-Moyano, S., Córdoba, M. G., Villalobos, M. C., Casquete, R., & Hernández, A. (2017). Combined effect of antagonistic yeast and modified atmosphere to control *Penicillium expansum* infection in sweet cherries cv. ambrunés. *International Journal of Food Microbiology*, 241, 276-282.

[97] Sharma, R. R., Singh, D., & Singh, R. (2009). Biological control of postharvest diseases of fruits and vegetables by microbial antagonists: A review. *Biological Control*, *50*, 205-221.

[98] Palmieri, D., Ianiri, G., Del Grosso, C., Barone, G., De Curtis, F.; Castoria, R., & Lima, G. (2022) Advances and Perspectives in the Use of Biocontrol Agents against Fungal Plant Diseases. *Horticulturae*, *8 (7)*, 577.

[99] Liu, J., Sui, Y., Wisniewski, M., Droby, S., & Liu, Y. (2013) Review: Utilization of Antagonistic Yeasts to Manage Postharvest Fungal Diseases of Fruit. *Int. J. Food Microbiol.*, *167* (2), 153-160.

[100] Muccilli, S., & Restuccia, C. (2015) Bioprotective Role of Yeasts. *Microorganisms*, *3* (4), 588-611.

[101] Pantelides, I. S., Christou, O., Tsolakidou, M. D., Tsaltas, D., & Ioannou, N. (2015). Isolation, identification and in vitro screening of grapevine yeasts for the control of black Aspergilli on grapes. *Biological Control*, *88*, 46–53.

[102] Di Francesco, A., Ugolini, L., Lazzeri, L., & Mari, M. (2015). Production of volatile organic compounds by *Aureobasidium pullulans* as a potential mechanism of action against postharvest fruit pathogens. *Biological Control*, *81*, 8-14.

[103] Parafati, L., Vitale, A., Restuccia, C., & Cirvilleri, G. (2017). Performance evaluation of volatile organic compounds by antagonistic yeasts immobilized on hydrogel spheres against gray, green and blue postharvest decays. *Food Microbiology*, 63, 191-198.

[104] Rodriguez Assaf, L. A., Pedrozo, L. P., Nally, M. C., Pesce, V. M., Toro, M. E., Castellanos de Figueroa, L. I., Vazquez, F. (2020) Use of Yeasts from Different Environments for the Control of *Penicillium expansum* on Table Grapes at Storage Temperature. *Surface*, *320*, 108520.

[105] Terao, D., de Lima Nechet, K., Ponte, M. S., de Holanda Nunes Maia, A., de Almeida Anjos, V. D., & de Almeida Halfeld-Vieira, B. (2017). Physical postharvest treatments combined with antagonistic yeast on the control of orange green mold. *Scientia Horticulturae*, *224*, 317-323.

[106] Wei, Y., Mao, S., & Tu, K. (2014). Effect of preharvest spraying *Cryptococcus laurentii* on postharvest decay and quality of strawberry. *Biological Control*, *73*, 68-74.

[107] Qin, X., Xiao, H., Xue, C., Yu, Z., Yang, R., Cai, Z.; & Si, L. (2015) Biocontrol of Gray Mold in Grapes with the Yeast *Hanseniaspora Uvarum* Alone and in Combination with Salicylic Acid or Sodium Bicarbonate. *Postharvest Biol. Technol.*, *100*, 160-167.

[108] [108] Zheng, X., Zheng, L., Xia, F., Li, J., Zhou, W., Yuan, L., Rao, S., & Yang, Z. (2023). Biological control of blue mold rot in apple by *Kluyveromyces marxianus* XZ1 and the possible mechanisms of action. *Postharvest Biology and Technology*, *196*, 112179.

[109] Prendes, L. P., Merín, M. G., Fontana, A. R., Bottini, R. A., Ramirez, M. L., & Morata de Ambrosini, V. I. (2018). Isolation, identification and selection of antagonistic yeast against *Alternaria alternata* infection and tenuazonic acid production in wine grapes from Argentina. *International Journal of Food Microbiology*, *266*, 14-20.

[110] Lemos Junior, W. J. F., Bovo, B., Nadai, C., Crosato, G., Carlot, M., Favaron, F., Giacomini, A., & Corich, V. (2016). Biocontrol ability and action mechanism of *Starmerella bacillaris* (Synonym *Candida zemplinina*) isolated from wine musts against gray mold disease agent *Botrytis cinerea* on grape and their effects on alcoholic fermentation. *Frontiers in Microbiology*, *7*, 1249.

[111] Makroo, H. A., Naqash, F., & Fayaz, S. (2021). Antimicrobial compounds to improve foods' shelf life. *Food Waste Recovery*, 461-482.

[112] Bartel, C., Roach, M., Onetto, C., Curtin, C., Varela, C., & Borneman, A. (2021). Adaptive evolution of sulfite tolerance in B. bruxellensis. *FEMS Yeast Research*, *21*(5), foab036.

[113] Chauhan, K., & Rao, A. (2024). Clean-label alternatives for food preservation: An emerging trend. *Heliyon*. 10(16, e35815.

[114] Spadaro, D., & Droby, S. (2015). Unraveling the mechanisms used by antagonistic yeast to control postharvest pathogens on fruit. In *III International Symposium on Postharvest Pathology: Using Science to Increase Food Availability 1144* (pp. 63-70).

[115] Mehlomakulu, N. N., Setati, M. E., & Divol, B. (2014). Characterization of novel killer toxins secreted by wine-related non-*Saccharomyces* yeasts and their action on *Brettanomyces* spp. *International Journal of Food Microbiology*, *188*, 83-91.

[116] Becker, B., & Schmitt, M. J. (2017). Yeast killer toxin K28: biology and unique strategy of host cell intoxication and killing.*Toxins*, *9*(10), 333.

[117] Belda, I., Ruiz, J., Alonso, A., Marquina, D., & Santos, A. (2017). The biology of *P. membranifaciens* killer toxins. *Toxins*, *9*(4), 112.

[118] He, Y., Degraeve, P., & Oulahal, N. (2024). Bioprotective yeasts: Potential to limit postharvest spoilage and to extend shelf life or improve microbial safety of processed foods. *Heliyon*, *10*(3). e24929.

[119] Contarino, R., Brighina, S., Fallico, B., Cirvilleri, G., Parafati, L., & Restuccia, C. (2019). Volatile organic compounds (VOCs) produced by biocontrol yeasts. *Food Microbiology*, *82*, 70-74.

[120] Lutz, M. C., Lopes, C. A., Rodriguez, M. E., Sosa, M. C., & Sangorrín, M. P. (2013). Efficacy and putative mode of action of native and commercial antagonistic yeasts against postharvest pathogens of pear. *International Journal of Food Microbiology*, *164*(2-3), 166-172.

[121] Czarnecka, M., Żarowska, B., Połomska, X., Restuccia, C., & Cirvilleri, G. (2019). Role of biocontrol yeasts *D. hansenii* and *W. anomalus* in plants' defence mechanisms against *Monilinia fructicola* in apple fruits. *Food Microbiology*, *83*, 1-8.

[122] Xu, N., Gao, H., Wang, Y., Liu, C., Hu, L., He, A., ... & Xin, F. (2024). Recent advances in bio-based production of organic acids by genetically engineered yeasts. *Biochemical Engineering Journal*, 109587.

[123] Karabegović, I., Malićanin, M., Danilović, B., Stanojević, J., Stojanović, S. S., Nikolić, N., & Lazić, M. (2021). Potential of non-*Saccharomyces* yeast for improving the aroma and sensory profile of Prokupac red wine. *OENO one*, *55*(2), 181-195.

[124] Muñoz-Redondo, J. M., Puertas, B., Cantos-Villar, E., Jiménez-Hierro, M. J., Carbú, M., Garrido, C., ... & Moreno-Rojas, J. M. (2021). Impact of sequential inoculation with the non-*Saccharomyces T. delbrueckii* and *M. pulcherrima* combined with *S. cerevisiae* strains on chemicals and sensory profile of rosé wines. *Journal of Agricultural and Food Chemistry*, *69*(5), 1598-1609.

[125] Sica, J., Vendramini, C., Nadai, C., Molinelli, Z., Carlot, M., Giacomini, A., & Corich, V. (2024). Strain prevalence and killer factor only partially influence the fermentation activity of pairwise *S. cerevisiae* wine strains inoculation. *Plos one*, *19*(4), e0300212.

[126] Banani, H., Spadaro, D., Zhang, D., Matic, S., Garibaldi, A., & Gullino, M. L. (2014). Biocontrol activity of an alkaline serine protease from *Aureobasidium pullulans* expressed in *Pichia pastoris* against four postharvest pathogens on apple. *International Journal of Food Microbiology*, *182*, 1-8.

[127] Ballet, N., Souche, J. L., & Vandekerckove, P. (2015, June). Efficacy of *C. oleophila,* strain O, in preventing postharvest diseases of fruits. In *III International Symposium on Postharvest Pathology: Using Science to Increase Food Availability 1144* (pp. 105-112).

Chapter 5

Biological Control in Tomato and Pepper Under Different Growing Conditions: The Potential for Disease Management of *Pseudomonas* Spp., *Trichoderma* Spp. and Microbial Consortia

N. A. Pastor[1,*]
J. A. Andrés[1,2]
V. Autrán[2]
A. M. Torres[3]
M. Rovera[3]
and L. B. Guiñazú[1,2]

[1]Instituto de Investigación en Micología y Micotoxicología (IMICO) - Consejo Nacional de Investigaciones Científicas y Técnicas (CONICET) – Departamento de Microbiología e Inmunología, Facultad de Ciencias Exactas, Físico-Químicas y Naturales, Universidad Nacional de Río Cuarto (UNRC), Río Cuarto, Córdoba, Argentina
[2]Departamento de Biología Agrícola, Facultad de Agronomía y Veterinaria, Río Cuarto, Córdoba, Argentina

Abstract

Biological control of plant pathogens offers a sustainable complement for reducing chemical pesticides in the management of bacterial and

[*] Corresponding Author's Email: npastor@exa.unrc.edu.ar

In: Biocontrol Agents (BCAs)
Editor: Daiana García
ISBN: 979-8-90134-102-5

fungal infections in commercially important crops such as tomato and pepper. The function of biocontrol agents, specifically *Pseudomonas* spp., *Trichoderma* spp., and microbial consortia, in reducing bacterial and fungal pathogen-induced diseases is the main topic of this chapter. The production of antimicrobial chemicals, competition for nutrients, and induction of systemic resistance in plants are key characteristics of *Pseudomonas* species, whereas some *Trichoderma* species can stimulate plant growth, synthesize enzymes, and/or display mycoparasitic activity. Microbial consortia are also investigated for their potential to offer more stable and complete disease control through synergistic effects. However, depending on the growing environment - hydroponics, growth chambers, greenhouses, and field systems - the effectiveness of biocontrol agents might differ greatly, each with its own set of difficulties. This chapter examines biological control methods that are currently studied for tomato and pepper crops, where the use of *Trichoderma* and *Pseudomonas* strains has been demonstrated to be successful. Furthermore, the application of microbial consortia - combinations of advantageous microorganisms - is emphasized as an effective approach to boost resilience against diseases, improve plant health, and increase biocontrol efficacy. The importance of further research to optimize the effectiveness and scalability of these strategies in diverse crop environments is also emphasized.

Keywords: tomato, pepper, biocontrol, beneficial microorganisms, microbial consortia

Introduction

The constant increase in food consumption has resulted in a reliance on agricultural chemicals that are harmful to human health. In addition to causing ecological imbalance, these chemical-based formulations also cause ecotoxicity. Plant diseases can be controlled by a variety of biocontrol techniques, such as those employed in organic agriculture, which are safe for the environment, reasonably priced, and have the potential to greatly increase plant yield (Pandit et al., 2022). Crop plants can interact with microorganisms to enhance their growth, development, and resistance to disease attack. Microbial metabolites are valued for their exceptional capacity to promote plant development and their effective biocontrol capabilities. By enhancing nutrient uptake, producing antagonistic compounds like antibiotics, hydrogen cyanide (HCN), volatile organic compounds (VOCs), and hydrolytic enzymes, as well as by producing plant growth regulators and siderophores, bio-complexes, such as biofertilizers and biopesticides, promote growth and shield plants from a variety of biotic and abiotic stressors (Elnahal et al., 2022).

This chapter examines recent research on biocontrol methods for protecting pepper and tomato crops. Several strains of *Trichoderma* and *Pseudomonas* have proven to be especially successful in inhibiting soil-borne diseases and promoting plant growth. Research shows that microbial consortia, carefully crafted groups of complementary advantageous microorganisms, have the potential to be a cutting-edge strategy for sustainable agriculture, going beyond single-strain applications. Through processes including nutrient solubilization, rhizosphere competition, and induced systemic resistance, synergistic consortia not only increase disease resistance but also enhance general plant health, satisfying the growing need for an environmentally conscious agriculture.

Pseudomonas Strains and Their Potential for Biological Control in Agriculture

Because of their remarkable capacity to inhibit a broad spectrum of crop-damaging pathogens, *Pseudomonas* species are crucial for the biological control of plant diseases. They are mostly isolated from plant rhizospheres and soil. *Pseudomonas* species can adapt to a range of environments due to their metabolic diversity. They could produce a variety of antimicrobial substances that prevent plant diseases from developing, thereby lowering their incidence. The vast body of research on *Pseudomonas* biocontrol agents illustrates how well they contribute to preventing diseases induced by bacteria, nematodes, fungi, and oomycetes. Several *P. chlororaphis* isolates have been the subject of substantial investigation, and several of these strains are marketed as biopesticides. With their ability to directly control a wide range of pathogens and create plant resistance, these strains demonstrate their versatility in disease management (Costa-Gutiérrez et al., 2022; Höfte, 2022). For example, Rovera et al. (2014) showed that *P. chlororaphis* subsp. *aurantiaca* SR1 can enhance the biomass of soybean plants and considerably decrease the damping-off of soybeans caused by *Macrophomina phaseolina*. The potential of *P. chlororaphis* subsp. *aurantiaca* SR1 to lessen the occurrence of damping-off in soybeans was based on its capacity to colonize soybean roots and produce pyrrolnitrin (PRN), phenazine-1-carboxylic acid (PCA), and hydrogen cyanide (HCN). Additionally, strains of *P. putida* have shown a great deal of promise in biocontrol applications, especially in the management of soil-borne diseases that impact on a variety of crops. By incorporating *P. putida* into biocontrol measures, it is highlighted that this beneficial bacterium in agriculture not only helps to manage diseases but also enhances plant health. The use of *P. putida* in commercial bioproducts is still restricted, despite its attractive capabilities observed in field tests (Costa-Gutiérrez et al., 2022; Höfte, 2022). The potential of *Pseudomonas* spp. to biologically control plant diseases is substantial. In this section, we focused on recent works that studied the biocontrol capacity of *Pseudomonas* strains for improving the health and growth of tomato and pepper.

Pseudomonas Strains for Biological Control of Pathogens That Attack Tomato

Suresh et al. (2021) isolated fluorescent pseudomonads (FPs) from the rhizosphere that demonstrated the production of amylase, protease, cellulase, pectinase, chitinase, gelatinase, siderophore, HCN, and indole acetic acid (IAA). The authors of the study underlined the importance of the structural integrity of the targeted pathogens' cell walls, which can be jeopardized by the hydrolytic enzymes produced by FPs. Furthermore, the study showed that FPs, particularly isolate *P. fluorescens* VSMKU3054, exhibited potent antagonistic qualities against a number of fungal pathogens, such as *R. solani*, *F. oxysporum,* and *S. rolfsii*, successfully limiting their development and dissemination. Additionally, the study showed that FPs have a great deal of potential to stimulate plant growth, as demonstrated by the longer roots and shoots of tomato seedlings treated with cell-free culture filtrate from VSMKU3054. Felipe et al. (2021) investigated three new strains of microorganisms: VMAP1, linked to *P. soli*, and VMA11p and VM05, closely related to *Bacillus velezensis*. These strains were chosen due to their ability to biocontrol *Xanthomonas vesicatoria*, which causes bacterial spot in tomato plants. The purpose of the study was to determine how well these strains controlled the disease in tomato plants cultivated in greenhouse pots. Bacterial suspensions were irrigated into the soil, and foliar sprays were administered using cell-free supernatants (CFSs). Applying the strains beforehand through soil irrigation considerably lessened the severity of the bacterial spot. However, using the CFSs via foliar application proved to be the most successful tactic. *P. fluorescens* VSMKU3054 was investigated further by Suresh et al. (2022), who mainly concentrated on its function as a biocontrol agent against *Ralstonia solanacearum*. One of the main findings of the study is that *P. fluorescens* VSMKU3054 effectively induces systemic resistance (ISR) in tomato plants, strengthening their defense mechanisms against the pathogen. In addition to an increase in the overall phenol content in the plants treated with the bacterium, this induction was typified by the activation of several defensive enzymes, such as phenylalanine ammonia-lyase (PAL), lipoxygenase (LOX), polyphenol oxidase (PPO), and peroxidase (POX). Ghadamgahi et al. (2022) concentrated on the plant growth-promoting rhizobacterium (PGPR) strain FG106, which was isolated from healthy nursery tomato plants. The biocontrol capability of this strain was assessed against tomato crop pathogens, with a focus on *Xanthomonas euvesicatoria* pv. *perforans* and *Clavibacter michiganensis* subsp. michiganensis, two important pathogens that cause bacterial cankers and spots in tomato. A soil drenching technique was used to inoculate tomato plants. The assays included three-week-old tomato seedlings that were kept in a controlled growth chamber using specific humidity and temperature settings. The results indicated that *P. aeruginosa* FG106 exhibited significant plant growth-promoting attributes and demonstrated high levels of inhibition against the growth and pathogenicity of the phytopathogens in the tested conditions. Mekonnen et al. (2022) investigated PGPR as biocontrol agents in a controlled greenhouse setting against the pathogen *R. solanacearum*. The bacteria being studied were *Pseudomonas* isolates BDUA2 and BDUA3, as well as *Bacillus* isolate BDUA1, which were obtained from the tomato rhizosphere. After being uprooted, the seedlings were immersed for 60 minutes in solutions of the chosen PGPR isolates (BDUA1, BDUA2, and BDUA3). The purpose of this soaking procedure was to improve the rhizobacteria's colonization of the plant roots prior to moving them into the pots with the potting soil. The incidence of bacterial wilt was considerably decreased by both BDUA1 and BDUA2. The utilization of beneficial *Pseudomonas* strains as biocontrol agents against *Agrobacterium rhizogenes*, the causative agent of hairy root disease (HRD) in tomatoes cultivated hydroponically, was the main emphasis of de Freitas & Taylor (2023). Among the microorganisms examined were *P. protegens* strain 1B1, *P. brassicacearum* strain 93G8, and *P. chlororaphis* strain 48G9, which showed notable *in vitro* activity against *A. rhizogenes*. They tested the efficacy of *Pseudomonas* strains in lowering disease incidence by treating plants in hydroponic systems with these beneficial bacteria. Plants treated with *Pseudomonas* strains 1B1 and 93G8 demonstrated a considerable decrease in disease incidence, whereas untreated plants had an almost 100% disease incidence. Also, Ajijah et al. (2023) studied *P. protegens* ML15, a plant growth-promoting bacterium isolated from the Błędowska Desert. The target pathogen was *Botrytis cinerea*, which is responsible for causing gray mold in postharvest tomato fruits. Fruits were treated with bacterial cultures, either the CFS of *P. protegens* ML15 or the culture suspension, as part of the application process. The tomatoes were inoculated with the pathogen *B. cinerea* in order to evaluate how well the treatments worked to prevent fungal growth and disease development. When compared to the control treatments four days after *B. cinerea* inoculation, the results showed that both *P. protegens* ML15 treatments considerably decreased fungal growth and alleviated disease lesions. Notably, there were no significant differences in disease lesion size and fungal growth between treatments, indicating that the pathogen was similarly inhibited by both treatments. Finally, Taha et al. (2023) investigated strains of *B. subtilis*, *B. amyloliquefaciens*, *P. resinovorans*, *P. alcaligenes*, *P. putida,* and *P. stutzeri* against *Geotrichum candidum* and *Alternaria alternata*, both of which are known to cause significant postharvest rots in tomato fruits. The bacterial bioagents were applied to tomato fruits in order to evaluate their antifungal effectiveness. The treated fruits were inoculated with the pathogens, and disease incidence and severity were measured over predetermined storage times to assess the bioagents' efficacy. According to the observations, *B. subtilis* and *P. stutzeri* were the most successful biocontrol agents, significantly lowering the incidence and severity of *G. candidum* disease. In addition to lowering the frequency of fruit rot, the use of these bioagents enhanced quality characteristics such as weight loss, vitamin C content, total soluble solids (TSS), and acidity during storage.

Key aspects from these recent works on *Pseudomonas* for biocontrol in tomato are compiled and summarized in Table 1.

Pseudomonas Strains for Biological Control of Pathogens That Attack Pepper

Hyder et al. (2021) examined a variety of rhizobacterial isolates, emphasizing the potential of *Pseudomonas* strains to control the damping-off disease that affects pepper plants and is caused by the soil-borne pathogen *Pythium myriotylum*. Before seeding, surface-sterilized seeds were soaked in bacterial suspensions as part of the application method. To provide a controlled environment and assess how well the bacterial isolates combated the disease, the soil was further treated with a *P. myriotylum* zoospore suspension. In comparison to untreated controls, the results showed that the administration of the bacterial isolates considerably increased seed germination rates and decreased seedling mortality. The study specifically noted higher vigor indices and increased root and shoot weights, as well as improved plant growth features. The studied bacterial isolates, including those belonging to the *Pseudomonas* genus, showed the capacity to generate beneficial compounds such as IAA, HCN, and siderophores, which support plant growth and inhibit pathogens. Furthermore, bacterization enhanced the activity of defense-related enzymes such as PAL, PPO, and POX under pathogen pressure.

To combat the pathogen *Phytophthora capsici*, Escobedo et al. (2022) used *Pseudomonas* species as part of a biological management strategy. Their study assessed the efficacy of consortia based on mycorrhizal fungi and rhizobacteria, with a focus on improving plant development and resistance to this pathogen. A subjective scale was used in the study to evaluate the degree of plant wilt and mortality, enabling a thorough examination of the pathogen's effects throughout time. According to the findings, applying *Pseudomonas* strains along with mycorrhizal fungi greatly enhanced plant development and decreased the prevalence of *P. capsici*-caused disease. Comparing the co-inoculation technique to other treatments, the study found that it decreased pathogen harm and increased seedling emergence. On the other hand, Jiménez-Pérez et al. (2023) investigated the antagonistic activity of a *P. donghuensis* strain against pepper phytopathogens, specifically focusing on *F. oxysporum* and *P. aphanidermatum*, which are known to induce "Damping off" in seedlings grown in greenhouse environments. In the experimental setup, culture plates supplemented with Potato Dextrose Agar (PDA) were used to inoculate the pathogens. *In vitro* confrontation assays against the target pathogens were used to evaluate the antagonistic effects of *P. donghuensis*. The findings showed that the growth of *P. aphanidermatum* was significantly inhibited by this bacterial strain. It had no effect on *F. oxysporum*, though, which kept growing quickly and filled the culture plate in less than 36 hours. Because of its antagonistic characteristics, the authors concluded that *P. donghuensis* is a promising candidate for biological control of *P. aphanidermatum* in pepper crops. The results additionally indicated that the management of "Damping off" diseases might be improved by combining *P. donghuensis* with another biocontrol agent, like *B. subtilis*. In 2023, Hussein et al. investigated a strain of *P. fluorescens*. The efficiency of this bacterium in controlling the target disease, *Leveillula taurica*, which causes pepper plants to develop powdery mildew, was assessed. Applying the bacterial cultures to the pepper plants was done by foliar spray.

Table 1. Examples of *Pseudomonas* strains recently proven to control pathogens that cause tomato diseases

Biocontrol agent(s)	Target pathogen(s)	Mechanism(s) of action	Efficacy	Conditions of the study	Advantage(s)	Reference
Pseudomonas fluorescens	*Ralstonia solanacearum, Rhizoctonia solani, Fusarium oxysporum, Sclerotium rolfsii*	Hydrolytic enzymes, hydrogen cyanide (HCN), and antibiosis	Significant suppression of fungal pathogens	*In vitro* assays	Also promotes early plant growth	Suresh et al. (2021)
Pseudomonas soli	*Xanthomonas vesicatoria*	Secondary metabolites, induction of systemic defense responses	Foliar application of supernatant highly reduced disease severity	Greenhouse. Inoculation of soil and leaves	Treatments did not negatively affect the growth of plants	Felipe et al. (2021)
Pseudomonas fluorescens	*Ralstonia solanacearum*	Induction of systemic resistance in tomato plants. Activation of various defense enzymes	Strong efficacy	*In vitro*	Plants resistant to entry and multiplication of *R. solanacearum*	Suresh et al. (2022)
Pseudomonas aeruginosa	*Alternaria alternata, Botrytis cinerea, Clavibacter michiganensis subsp. michiganensis, Phytophthora colocasiae, P. infestans, Rhizoctonia solani, and Xanthomonas euvesicatoria pv. perforans.*	Hydrolytic enzymes, HCN siderophores, VOCs	Significantly reduced disease severity	Controlled conditions. Greenhouse	Increased root and shoot weights	Ghadamgahi et al. (2022)

Biocontrol agent(s)	Target pathogen(s)	Mechanism(s) of action	Efficacy	Conditions of the study	Advantage(s)	Reference
Pseudomonas sp.	*Ralstonia solanacearum*	Hydrolytic enzymes, HCN, siderophores	Significant reduction in the incidence of bacterial wilt	Greenhouse conditions	Dual role: growth promotion and disease control	Mekonnen et al. (2022)
P. protegens, P. clororaphis, P. brassicacearum	*Agrobacterium rhizogenes*	Competition. Induction of systemic defense responses	Up to 80% reduction in disease incidence	*In vitro*. Hydroponic conditions	Reduction in the reliance on chemical treatments	de Freitas & Taylor (2023)
Pseudomonas protegens	*Botrytis cinerea*	VOCs, antifungal compounds, induction of defense responses	Culture suspension and cell-free supernatant mitigated disease development	Fruits under controlled conditions	Many modes of action offer a strong postharvest disease management approach.	Ajijah et al. (2023)
Pseudomonas stutzeri, Pseudomonas alcaligenes	*Geotrichum candidum, Alternaria alternata*	Hydrolytic enzymes	High reduction in disease incidence and severity	*In vitro* conditions. Fruits under controlled conditions	Improvement of the physiological traits of tomato fruits during storage	Taha et al. (2023)

Table 2. Examples of *Pseudomonas* strains recently proven to control pathogens that cause pepper diseases

Biocontrol agent(s)	Target pathogen(s)	Mechanism(s) of action	Efficacy	Conditions of the study	Advantages	Reference
Pseudomonas putida, Pseudomonas libanensis	*Pythium myriotylum*	HCN, increase of plant defenses, siderophores	Significant reductions in seedling mortality rates	Controlled pot experiments	Ability to enhance plant growth while suppressing the target pathogen	Hyder et al. (2021)
Pseudomonas lini, Pseudomonas fluorescens	*Phytophthora capsici*	None specifically suggested	Healthier plants with fewer symptoms of disease	Greenhouse. Controlled conditions	Plant growth promotion	Escobedo et al. (2022)
Pseudomonas donghuensis	*Pythium aphanidermatum*	Siderophores	It may need to be used in combination with other antagonists for broader efficacy	Controlled conditions *in vitro*	High capacity for siderophore production	Jiménez-Pérez et al. (2023)
Pseudomonas fluorescens	*Leveillula taurica*	Enhancement of the plant's defense responses	Moderate efficacy in managing the disease	Greenhouse conditions	Does not leave harmful residues on crops	Hussein et al. (2023)
Pseudomonas aeruginosa	*Phytophthora capsici*	Production of secondary metabolites, induction of systemic resistance	High efficacy	Controlled settings. Field assays	The strain also promotes increased plant height and yield	Li et al. (2024)

The authors reported that their strain of *P. fluorescens* demonstrated a moderate level of efficiency in controlling the disease. Furthermore, the use of *P. fluorescens* and other bacterial strains led to a significant increase in pepper plant yield, indicating that it can boost plant productivity while confronting the pathogen. According to this study's findings, *P. fluorescens* not only lessened the severity of the disease but also enhanced crop yield, indicating that it could be a viable substitute for chemical fungicides.

Finally, *P. aeruginosa* strain Pa608, which was isolated from the rhizosphere of pepper plants, was the subject of a study by Li et al. (2024). This strain was found to be a biocontrol agent against *P. capsici*, the target pathogen that causes severe pepper blight. It was demonstrated that the strain successfully colonized the pepper roots. Because it enables the strain to directly exert its positive effects in the rhizosphere, this colonization is essential to the biocontrol process. Both *in vitro* and *in vivo*, the results showed that *P. aeruginosa* Pa608 effectively suppressed mycelial development and had strong antagonistic activity against *P. capsici*. The field tests and potted studies demonstrated a strong control efficacy in lowering the pepper blight disease index. Additionally, it was discovered that strain Pa608 stimulated plant growth, as shown by higher yield and plant height.

Key aspects from these recent works on *Pseudomonas* for biocontrol in pepper are compiled and summarized in Table 2.

Pathogens and Mechanisms of Biocontrol by *Pseudomonas* Spp

In Figure 1, we intended to summarize the mechanisms mentioned by these authors that are used by *Pseudomonas* strains to control pathogens. By targeting the roots and vascular system, *R. solanacearum* induces bacterial wilt, which results in fast withering. Root rot and damping-off are caused by *R. solani*, which harms the roots and the base of the stem. By causing *Fusarium* wilt, *F. oxysporum* clogs the vascular system. Southern blight, or rotting stems close to the soil, is produced by *S. rolfsii*. *X. vesicatoria* and *X. euvesicatoria* pv. *perforans* are the causative agents of bacterial spot infections that impact fruits and foliage. Bacterial canker caused by *C. michiganensis* subsp. *michiganensis* damages fruits, leaves, and stems. Root growth is hindered by *A. rhizogenes*, which produces hairy root disease. Fungal diseases that infect leaves, stems, and fruits include *B. cinerea* (gray mold) and *A. alternata* (blight), while *G. candidum* produces sour rot in ripe fruits. On the other hand, pepper plants are vulnerable to several destructive pathogens. We reviewed works dealing with those mentioned next. Root rot and damping-off are caused by *P. myriotylum* and *P. aphanidermatum*, which significantly reduce seedling survival. Fruit, stem, and root rot are caused by blight, which is induced by *P. capsici*, one of the most virulent pathogens. Lastly, *L. taurica* reduces yield by causing powdery mildew, which covers leaves with white fungal growth.

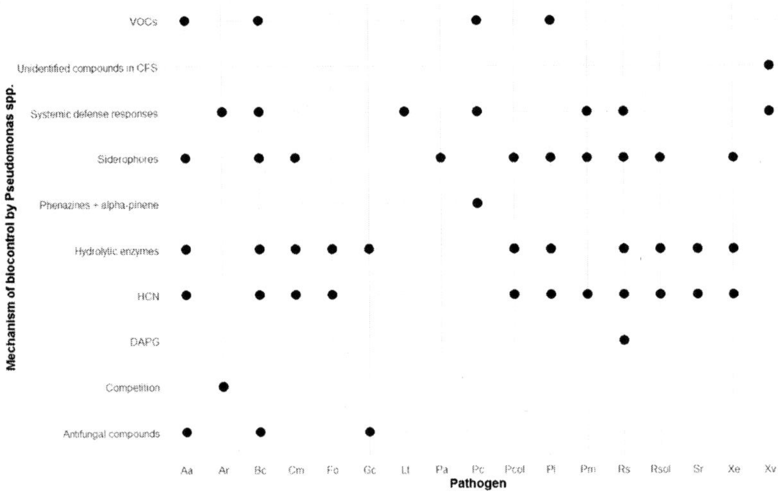

Figure 1. Summary of mechanisms that can be used by *Pseudomonas* strains to control tomato and pepper pathogens. VOCs = Volatile Organic Compounds; CFS = Cell Free Supernatant; HCN = Hydrogen Cyanide; DAPG = 2,4-Diacetylphloroglucinol; Aa = *Alternaria alternata*; Ar = *Agrobacterium rhizogenes*, Bc = *Botrytis cinerea*; Cm = *Clavibacter michiganensis* subsp. *michiganensis*; Fo = *Fusarium oxysporum*; Gc = *Geotrichum candidum*; Lt = *Leveillula taurica*; Pa = *Pythium aphanidermatum*; Pc = *Phytophthora capsici*; Pcol = *Phytophthora colocasiae;* Pi = *Phytophthora infestans;* Pm = *Pythium myriotylum*; Rs = *Ralstonia solanacearum*; Rsol = *Rhizoctonia solani*; Sr = *Sclerotium rolfsii*; Xe = *Xanthomonas euvesicatoria* pv. *perforans*; Xv = *Xanthomonas vesicatoria*.

Trichoderma Strains and Their Potential for Biological Control in Agriculture: Formulations for Tomato and Pepper

The use of microorganisms capable of exercising biological control against other species represents a sustainable ecological alternative for disease management. Some advantages of their use are versatility to act against more than one pathogen in different crops, promotion of plant growth, induction of plant defenses against attacks by pathogens, and improvement and maintenance of soil productivity (Hermosa et al., 2013). Among the most used biocontrol agents, we find the genus *Trichoderma* (Benítez et al., 2004).

The success of *Trichoderma* strains as biocontrol agents is due to their high reproductive capacity, ability to survive under very unfavorable conditions, efficiency in the utilization of nutrients, capacity to modify the rhizosphere, strong aggressiveness against phytopathogenic fungi, and efficiency in promoting plant growth and defense mechanisms. These properties have made *Trichoderma* a ubiquitous genus present in any habitat and at high population densities (Chet et al., 1997). The genus *Trichoderma* (teleomorph *Hypocrea*) comprises several filamentous, avirulent, and symbiotic fungal species commonly found in the soil of the most diverse ecosystems (Harman et al., 2004; Hermosa et al., 2012). Species within *Trichoderma* are known worldwide for their antagonistic abilities against plant pathogens since they reduce the severity of diseases through their effective mycoparasitic activity and their high reproductive and competitive capacity (Lopes et al., 2012; Erazo et al., 2021). Several mechanisms have been reported to contribute to the ability of *Trichoderma* to inhibit pathogenic microorganisms. These include its high capacity to degrade various substrates, which results in rapid growth. This may limit the growth of other microorganisms due to competition for organic nutrients, light, minerals, and space (Howell, 2003; Naher et al., 2014; Sharma et al., 2012; Stewart and Hill, 2014). Other mechanisms include the production of secondary metabolites with antibiotic activity (Maruyama et al., 2020; Kovács et al., 2021) and lytic enzymes primarily against the cell walls of other organisms that it usually attacks by a process of mycoparasitism (Druzhinina et al., 2018; Erazo et al., 2021). It has also been identified as an inducer of systemic resistance in plants (Saravanakumar et al., 2016).

Tomato (*Solanum lycopersicum* L.) is one of the most important crops worldwide (Orus, 2023). This plant belongs to the Solanaceae family and is native to the Andean region of Ecuador, Peru, and Bolivia (Peralta and Sponer, 2007; Bima and Fontán, 2020). It is widely adapted to different agroclimatic conditions, from the tropics to temperate regions. This vegetable is of great importance worldwide, given that it is one of the most consumed. It is marketed in the food industry as fresh or processed tomatoes, with the former being continuously supplied. Its fruit is marketed for consumption fresh, cooked, and/or for processing as canned peeled tomatoes, sauces, juices, and concentrates (Franco, 2005). Tomato fruits are rich in minerals, vitamins, essential amino acids, sugars, and fiber. In addition to their high fructose and glucose content, they are an important source of vitamins A and C, and contain high concentrations of iron and phosphorus. Furthermore, tomato-based products are considered highly valuable in diets due to their lycopene and flavonoid content (Lenucci et al., 2006). It is also used as a model plant for studies related to fruit quality and tolerance to biotic and abiotic stresses (Panthee and Chen, 2010).

Pepper (*Capsicum annuum* L.) is a plant of the *Solanaceae* family, native to Central and South America. The greatest economic importance of this crop originates in the commercialization of its fruits, which are rich in provitamin A, vitamin B, vitamin C, and minerals such as calcium, phosphorus, potassium, and iron (Maboko et al., 2012). It is consumed fresh, cooked, or as a condiment or "spice" in typical foods of various countries, and also in a range of industrial products (Loayza, 2001).

The use of *Trichoderma* as a biological control agent offers numerous advantages. It is a sustainable alternative to chemical fungicides, does not generate resistance in pathogens, is safe for people, plants, and the environment, promotes plant growth and health, and can be applied in a variety of ways. The first step in the biocontrol study is the identification of promising biocontrol agents. Once the biocontrol agent is identified and proven effective against plant pathogens over several reproducible results, the method of mass production, formulation, and application should be taken into consideration to stabilize the product during storage and to facilitate its delivery to the plant. Both solid and liquid formulations are used to produce suitable quantities of active and viable inocula of *Trichoderma* (Cumagun, 2014). Many biopesticide and biofertilizer products are now available on the market, the majority of which are based on beneficial symbionts of the genus *Trichoderma* (Woo et al., 2006).

Trichoderma spp., particularly *Trichoderma harzianum*, is widely employed in various formulations for the biocontrol of diseases in tomato crops. These formulations aim to deliver the beneficial fungus effectively to the plant and its rhizosphere, promoting growth and suppressing pathogens. Here, we include some methods and common formulations used for *Trichoderma* for the biocontrol of tomato and pepper:

- Seed treatment: procedures in which the conidia of *Trichoderma* are coated onto tomato seeds using adhesives (Taylor et al., 1994; Harman, 2006).
- Colonized Substrates: in which *Trichoderma* is grown on organic substrates like compost, bran, or grains, and this colonized material is

then incorporated into the soil (Whipps, 2001; dos Santos Pereira et al., 2021).

- Pellets: compressed formulations of *Trichoderma* biomass, sometimes combined with nutrients or other beneficial agents, for targeted application (Ramanujam et al., 2010; Abdulla, 2019).
- Wettable Powders: these formulations consist of conidia or mycelial fragments of *Trichoderma* combined with inert carriers (like talc or clay) and wetting agents to ensure proper dispersion in water for application as a soil drench or spray (Harman et al., 2004).
- Granules: *Trichoderma* propagules are incorporated into granular carriers, allowing for easier and more targeted application to the soil, often at planting or transplanting (Jeyarajan, 2006).
- Spore Suspensions: concentrated suspensions of *Trichoderma* conidia in a liquid medium, often with surfactants or sticking agents, for application as a soil drench, root dip, or foliar spray (Jackson et al., 1991).
- Microencapsulation: encapsulating *Trichoderma* propagules in protective matrices, such as alginate, to enhance their survival and provide controlled release in the soil (Muñoz-Celaya et al., 2012).

The choice of formulation depends on the target disease, application method, environmental conditions, and the specific needs of the tomato crop. Researchers are continuously exploring new and improved formulations to enhance the efficacy and stability of *Trichoderma* as a vital tool in sustainable tomato production.

Effectiveness of the Biological Control Agents *Pseudomonas* Spp. and *Trichoderma Harzianum* in Different Growing Environments

Biological control has gained importance in sustainable agriculture due to its ability to reduce dependence on synthetic chemicals and its potential to improve soil health and crop quality (Sharma et al., 2020). *Pseudomonas* spp. and *T. harzianum* are widely used as biocontrol agents due to their antagonistic properties against phytopathogens, which include the production of antifungal metabolites, competition for nutrients and space, and the induction of systemic resistance in plants (Harman et al., 2004; Compant et al., 2005; Pastor et al.,

2012; Mahapatra et al., 2024). However, the effectiveness of these microorganisms is influenced by factors such as crop type, environmental conditions, and the production system used.

Hydroponics

Hydroponic systems are a form of soilless cultivation that uses nutrient solutions to meet the mineral requirements of plants. They are implemented in controlled environments, such as greenhouses, allowing for food production in less favorable areas. Among the many agronomic advantages of this type of production are the wide variety of plant species that can be grown, the efficient use and recirculation of water and nutrients, the precise control of environmental variables, and the prevention of soil-borne diseases and pests (Lommen, 2007; Molitor, 1990). It is important to mention that the demand for hydroponic crops such as lettuce, tomatoes, and peppers has been increasing in our country and around the world (Argentine Hydroponic Association, 2025; Morning AgClips, 2024). Given the high nutrient concentrations in the solutions used, the growth of pathogenic microorganisms such as *Fusarium, Phytophthora*, and *Pythium* is common, causing serious losses for producers. Among the methods used to prevent contamination of hydroponic systems, the use of plant growth-promoting rhizobacteria (PGPR) of the genera *Pseudomonas* and *Trichoderma* stands out (Lee and Lee, 2015). Numerous strains of *Pseudomonas* sp. have shown high efficacy in suppressing root pathogens, such as *A. rhizogenes*, due to their ability to produce antifungal metabolites and increase plant growth (Chatterton et al., 2004; Renault et al., 2007; Lee et al., 2010; de Freitas and Taylor, 2023). The application of *Pseudomonas* spp. has shown high efficacy in suppressing root pathogens, such as *Pythium* spp. and *Phytophthora* spp., due to its ability to colonize roots and produce antifungal metabolites that inhibit the growth of these fungi (Amaradaşa et al., 2024). *T. harzianum* has also been evaluated in hydroponic systems, where its ability to colonize roots and produce hydrolytic enzymes has resulted in a significant reduction in the incidence of root diseases (López et al., 2010). Treatment with *T. harzianum* and *T. atroviride* has also been shown to stimulate growth and production of cucumbers and tomatoes (germination rate, dry weight, shoot length, and leaf area) (Yedidia et al., 2001).

Growth Chambers

Under controlled growth chamber conditions, *T. harzianum* has been observed to exhibit a high rate of rhizosphere colonization and increase plant biomass through the production of phytohormones such as indoleacetic acid (IAA) (Contreras-Cornejo et al., 2009). Furthermore, this microorganism can modulate the expression of plant defense-related genes, increasing their resistance to diseases (Vinale et al., 2008). Different strains of *Pseudomonas* spp. were able to improve the resistance of tomato plants by producing siderophores and volatile compounds that stimulate plant growth and limit the proliferation of pathogens such as *R. solani, A. alternata, B. cinerea,* and *P. infestans* (Raaijmakers et al., 2009; Ghadamgahi et al., 2022). In some studies, it has been reported that the combination of various strains of *Pseudomonas* spp. and *Trichoderma* spp. can enhance their biocontrol effects on pathogens such as *F. oxisporum, R. solani, P. colocasiae, P. capsici*, and *Pithium* spp., in pepper (Jena, 2012; Chemeltorit et al., 2017).

Greenhouses

In greenhouse crops, which provide more controlled conditions than open fields but where pathogen pressure is high due to controlled humidity and temperature, *Pseudomonas* spp. and *Trichoderma* spp. have proven effective in reducing diseases caused by phytopathogenic fungi, such as *R. solani, B. cinerea,* and *Scleritinia sclerotiorum* in pepper, tomato, and ornamental plants (Harman et al., 2004; Rini and Sulochana, 2007; Wang et al., 2021; Ampntelnour et al., 2024). Native strains of *Trichoderma* spp., *P. fluorescens,* and *B. subtilis* have been successfully used as biocontrol agents to control bacterial wilt caused by *Ralstonia* spp., one of the main pathogens of tomato, and the combined application of these beneficial microorganisms can offer even better results (Yendyo et al., 2018). However, the interaction with other soil microorganisms can influence their performance, so it is recommended to adjust the doses and application times according to the type of crop and the greenhouse conditions.

Field Conditions

Pathogens such as *Fusarium* spp. and *Phytophthora* spp. are causing significant losses in tomato crops, both in the field and in greenhouses, and remain significant limiting factors for horticultural production (Sing et al., 2017). The dual role of *Pseudomonas* as a biological control agent and plant growth promoter is a valuable tool for sustainable agriculture (Patel et al., 2020). Studies conducted with *P. aeruginosa* Pa608 on pepper plants have demonstrated antagonistic activity against *P. capsici*, effectively suppressing mycelial growth in greenhouses and in the field (Li et al., 2024). Under field conditions, the efficacy of these biocontrol agents can be affected by environmental factors such as temperature and soil moisture. Despite this, cases of successful pathogen suppression have been reported with the combined application of *Pseudomonas* sp. and *T. harzianum* (Rana et al., 2025). Isolations of *Trichoderma* spp. and *P. fluorescens* have been shown to be effective in controlling early blight of tomatoes, both in greenhouses and in the field, and have minimized yield losses caused by the disease (Kulimushi et al., 2021). Repeated application of these microorganisms in the soil has been observed to improve their establishment and persistence, increasing their long-term efficacy. Furthermore, the use of these biocontrollers in combination with appropriate agronomic practices, such as crop rotation and organic soil management, can enhance their beneficial effects. The integration of *Pseudomonas* sp. and *T. harzianum* into agricultural systems can be a key strategy for reducing the use of chemical pesticides, increasing crop resilience to pathogens and environmental stresses. However, further research into the interaction of these agents with different soil types, crops, and management conditions is essential to maximize their effectiveness and applicability in diverse production systems.

Microbial Consortia: An Approach for Enhanced Biocontrol of Diseases in Tomato and Pepper

This section highlights recent works in synergistic microorganism applications for plant pathogen suppression. According to Sood et al. (2020), inoculation using a combination of *B. subtilis* and *T. viride* showed some noteworthy advantages in controlling *R. solanacearum*-induced bacterial wilt disease in tomato plants. The consortium's effectiveness in preventing bacterial wilt was demonstrated by the total suppression of disease incidence upon application, which was one of the main benefits noted. A great increase in fresh plant weight and height was likewise observed in the consortium treatment, indicating its beneficial effects on plant growth. Furthermore, it was demonstrated that using the consortium was more successful than using the biocontrol agents alone. Tomato plants were inoculated with a microbial consortium by Pothiraj et al. (2021), which offered multiple advantages for controlling *Fusarium* wilt. *B. subtilis*, *P. putida*, *T. harzianum,* and *Chaetomium globosum* were part of the consortium. The purpose of the study was to confirm that the consortium could effectively control this disease in the field. With a 60% mean reduction in wilt incidence compared to the untreated control, the strain mixture outperformed the fungicide treatment and the untreated control. In their study, Rokni et al. (2021) examined a consortium of fungal bioinoculants, *T. simmonsii* and *Serendipita indica*, to increase their effect against *P. capsici*. Compared to individual treatments, the consortium's use resulted in a 68.4% decrease in the disease index, demonstrating a high protective impact against the pathogen. Significant pepper plant growth was also encouraged by this combination. Improved growth metrics, including stem and root length, fresh and dry weight, and stem crown breadth, demonstrated this. Additionally, in order to control *Colletotrichum truncatum*, which is known to induce anthracnose in peppers, Yadav et al. (2021) investigated combinations of *T. harzianum*, *T. asperellum,* and *Pacnibacillus dendritiformis*. Since this disease causes large crop losses, finding efficient control methods is essential. It was demonstrated that pepper plants developed systemic resistance to *C. truncatum* when exposed to the tested consortia of biocontrol agents. This indicates that after inoculation, plants grow more resistant to diseases. Inoculation with the consortium led to increased activities of various defense-related enzymes. This increase in enzyme activity strengthens the plant's defenses against the pathogen and aids in its degradation. The highest level of disease protection was offered by *T. asperellum* + *T. harzianum*. Phenolic chemicals, which are known to contribute to plant defense against diseases, also accumulated because of the inoculation. For improved biocontrol of *B. cinerea* in tomatoes, Li et al. (2023) investigated a consortium based on the combination of two biocontrol fungal agents, *T. harzianum* and *Clonostachys rosea*. Using *T. harzianum* and *C. rosea* together was more effective than using either agent alone in controlling *B. cinerea*. *In vitro* tests and field applications proved this, as the consortium exhibited strong fungicidal activities against the pathogen. Additionally, the authors discovered that the combination of these two biocontrol agents considerably stimulated tomato plant growth. This implies that the consortium promotes general plant health in addition to aiding in disease management. The authors observed that *T. harzianum* and *C. rosea* together caused tomato plants to develop resistance mechanisms against *B. cinerea* via the salicylic acid signaling pathway, which is essential for boosting the plant's defensive mechanisms. Jia et al. (2023) concentrated their investigation on *A. solani*, a noteworthy pathogen that causes tomato plant early blight disease. Due to the pathogen's damage to chloroplasts, affected plants may experience up to 90% reductions in yield. *B. subtilis* and *P. fluorescens*, two beneficial bacteria, were combined, and this consortium demonstrated a high inhibition rate against *A. solani*, lowering the disease incidence considerably in comparison to plants that were not treated. Antioxidant enzyme activity (peroxidase, catalase, and superoxide dismutase) significantly decreased after inoculation. This decrease implies that the consortium assisted in reducing tomato plant oxidative damage. Along with protecting against the disease, the consortium increased tomato biomass. The general productivity and health of the plant depend on this growth boost. Figure 2 covers the composition of the mentioned consortia tested for biocontrol of tomato and pepper diseases using a heatmap.

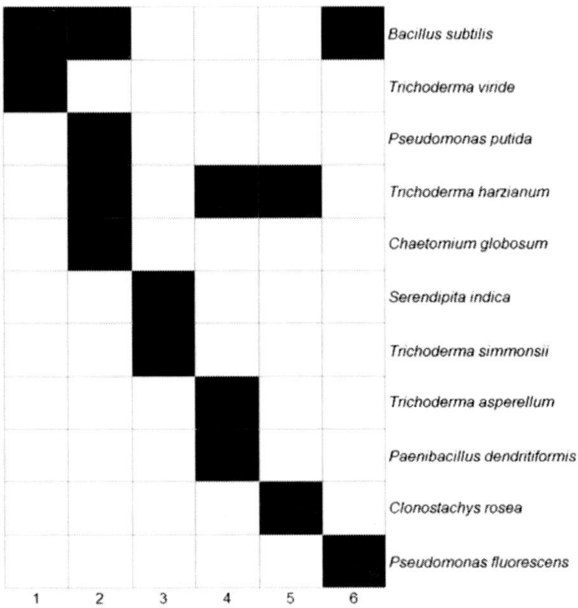

Figure 2. Heatmap comprising the microbial consortia reviewed in this section for control of tomato and pepper pathogens. Columns represent the consortia and rows the individual species. Dark squares represent combinations.

Conclusion

Inoculating plants with beneficial microorganisms, such as strains of *Pseudomonas* and *Trichoderma*, has several benefits, as we examined in this chapter. These benefits include enhancing resistance to diseases, promoting early growth, and increasing growth parameters in a variety of scenarios. In addition to promoting plant growth, these microorganisms also function as biological control agents, which lessens the need for chemical treatments and keeps hazardous residues from building up in horticultural products. They can even improve the physiological characteristics of fruits, contributing to a more sustainable agriculture. The effectiveness of *Pseudomonas* spp. and *Trichoderma* spp. varies depending on the growing environment, being more predictable in controlled systems such as hydroponics and growth chambers, where conditions are optimal for their establishment. When used in consortia, these microorganisms are far more successful than when used alone because of their synergistic impact, which enhances plant development and disease management. This combination approach to disease control for horticulture crops is more effective and dependable.

Acknowledgments

This work was supported by grants from Agencia Nacional de Promoción Científica y Tecnológica (MINCyT) through PICT2021-0963-Préstamo BID and Secretaría de Ciencia y Técnica de la Universidad Nacional de Río Cuarto (SECyT-UNRC).

References

Abdullah, F. (2019). Pellet formulation of *Trichoderma* isolated from aceh with potential for biocontrol of *Phytophthora* sp. ON. *CABI Digital Library* 3472322.

Ajijah, N., Fiodor, A., Dziurzynski, M., Stasiuk, R., Pawlowska, J., Dziewit, L., & Pranaw, K. (2023). Biocontrol potential of *Pseudomonas protegens* ML15 against *Botrytis cinerea* causing gray mold on postharvest tomato (*Solanum lycopersicum* var. *cerasiforme*). *Frontiers in Plant Science, 14*. https://doi.org/10.3389/fpls.2023.1288408.

Amaradasa, B. S., Kay, R. A., & Larkin, R. P. (2024). A root-colonizing endophyte, *Pseudomonas* IALR1619, offers protection to lettuce and cucumber against *Pythium ultimum* in hydroponic culture. *Frontiers in Microbiology*, 15, 1333846. https://doi.org/10.3389/fmicb.2024.1333846.

Ampntelnour, L., Poulaki, E. G., Dimitrakas, V., Mavrommati, M., Amourgis, G. G., & Tjamos, S. E. (2024). Enhancing *Botrytis* disease management in tomato plants: insights from a *Pseudomonas putida* strain with biocontrol activity. *Journal Of Applied Microbiology*, *135*(4). https://doi.org/10.1093/jambio/lxae094.

Asociación Hidropónica Argentina. *Potenciando la actividad en el País*. (s/f). Com.ar. Last accessed may 26th 2025, de https://asociacionhidroponica.com.ar/.

Benítez, T., Rincón, A. M., Limón, Antonio, M. C., & Codón, C. (2004). Biocontrol mechanisms of *Trichoderma* strains. *International Microbiology* 7, 249-260.

Bima P., & Fontán, H. (2020). Sistemas de producción de cultivos intensivos. Tomo II. Facultad de Ciencias Agropecuarias, Universidad Nacional de Córdoba.

Chatterton, S., Sutton, J. C., & Boland, G. J. (2004). Timing *Pseudomonas chlororaphis* applications to control *Pythium aphanidermatum*, *Pythium dissotocum*, and root rot in hydroponic peppers. *Biological Control: Theory and Applications in Pest Management*, *30*(2), 360–373. https://doi.org/10.1016/j.biocontrol.2003.11.001.

Chemeltorit, P. P., Mutaqin, K. H., & Widodo, W. (2017). Combining *Trichoderma hamatum* THSW13 and *Pseudomonas aeruginosa* BJ10–86: a synergistic chili pepper seed treatment for *Phytophthora capsici* infested soil. *European Journal of Plant Pathology*, *147*(1), 157–166. https://doi.org/10.1007/s10658-016-0988-5.

Chen, L. H., Cui, Y. Q., Yang, X. M., Zhao, D. K., & Shen, Q. R. (2012). An antifungal compound from *Trichoderma harzianum* SQR-T037 effectively controls *Fusarium* wilt of cucumber in continuously cropped soil. *Australasian Plant Pathology 41*(3), 239–245. https://doi.org/10.1007/s13313-012-0119-5.

Chet, I., Inbar, J., & Hadar, I. (1997). Fungal antagonists and mycoparasites. In: Wicklow DT, Söderström B (eds) The Mycota IV: Environmental and microbial relationships. Springer-Verlag, Berlin, pp. 165-184.

Compant, S., Duffy, B., Nowak, J., Clément, C., & Barka, E. A. (2005). Use of plant growth-promoting bacteria for biocontrol of plant diseases: principles, mechanisms of action, and future prospects. *Applied and Environmental Microbiology*, *71*(9), 4951–4959. https://doi.org/10.1128/AEM.71.9.4951-4959.2005.

Contreras-Cornejo, H. A., Macías-Rodríguez, L., Cortés-Penagos, C., & López-Bucio, J. (2009). *Trichoderma virens*, a plant beneficial fungus, enhances biomass production and promotes lateral root growth through an auxin-dependent mechanism in *Arabidopsis*. *Plant Physiology, 149*(3), 1579–1592. https://doi.org/10.1104/pp.108.130369.

Costa-Gutiérrez, S. B., Adler, C., Espinosa-Urgel, M., y de Cristóbal, R. E. (2022). *Pseudomonas putida* and its close relatives: Mixing and mastering the perfect tune for plants. *Applied Microbiology and Biotechnology*, *106*(9), 3351–3367. https://doi.org/10.1007/s00253-022-11881-7.

Cumagun, C. J. R. (2014). Advances in formulation of *Trichoderma* for biocontrol. In: Gupta, V. K., Schmoll, M., Herrera-Estrella, A., Upadhyay, RS, Druzhinina, I., & Tuohy M. (Eds). Biotechnology and Biology of *Trichoderma*, Elsevier, Ámsterdam, 527-532. https://doi.org/10.1016/b978-0-444-59576-8.00039-4.

de Freitas, C. C., & Taylor, C. G. (2023). Biological control of hairy root disease using beneficial *Pseudomonas* strains. *Biological Control: Theory and Applications in Pest Management*, *177*(105098), 105098. https://doi.org/10.1016/j.biocontrol.2022.105098.

dos Santos Pereira, T., Monteiro de Paula, A., Ferrari, L. H., da Silva, J., Borges Pinheiro, J., Navas Cajamarca, S. M., Jindo, K., Pupo Santos, M., Zandonadi, D. B., & Busato, J. G. (2021). *Trichoderma*-enriched vermicompost extracts reduce nematode biotic stress in tomato and bell pepper crops. *Agronomy* 11, 1655. https://doi.org/10.3390/agronomy11081655.

Druzhinina, I. S., Chenthamara, K., Zhang, J., Atanasova, L., Yang, D., Miao, Y., Rahimi, M. J., Grujic, M., Cai, F., Pourmehdi, S., Salim, K. A., Pretzer, C., Kopchinskiy, A. G., Henrissat, B., Kuo, A., Hundley, H., Wang, M., Aerts, A., Salamov, A., Lipzen, A., LaButti, K., Barry, K., Grigoriev, I. V., Shen, Q., & Kubicek, C. P. (2018). Massive lateral transfer of genes encoding plant cell wall-degrading enzymes to the mycoparasitic fungus *Trichoderma* from its plant-associated hosts. *PLoS Genetics* 14, e1007322. doi: 10.1371/journal.pgen.1007322.

Elnahal, A. S. M., El-Saadony, M. T., Saad, A. M. et al. (2022). The use of microbial inoculants for biological control, plant growth promotion, and sustainable agriculture: A review. *European Journal of Plant Pathology 162*, 759–792. https://doi.org/10.1007/s10658-021-02393-7.

Erazo, J. G., Palacios, S. A., Pastor, N., Giordano, F. D., Rovera, M., Reynoso, M. M., Venisse, J. S., & Torres, A. M. (2021). Biocontrol mechanisms of *Trichoderma harzianum* ITEM 3636 against peanut brown root rot caused by *Fusarium solani* RC 386. *Biological Control* 164, 104774. https://doi.org/10.1016/j.biocontrol.2021.104774.

Escobedo, L. L., Carrillo, M. G., Martínez, J. D., García, A. V., Castro, A. A., Rangel, P. P., & Puente, E. O. R. (2022). Consorcios de hongos micorrízicos y rizobacterias en el control biológico de *Phythopthora capsici* en cultivares de *Capsicum annuum*. *Biotecnia*, *24*(1), Article 1. https://doi.org/10.18633/biotecnia.v24i1.1490.

Felipe, V., Bianco, M. I., Terrestre, M., Mielnichuk, N., Romero, A. M., & Yaryura, P. M. (2021). Biocontrol of tomato bacterial spot by novel *Bacillus* and *Pseudomonas* strains. *European Journal of Plant Pathology*, *160*(4), 935–948. https://doi.org/10.1007/s10658-021-02297-6.

Franco, D. (2005). Tomate industrializado. Alimentos Argentinos, pp.1-31. Available in: http://www.alimentosargentinos.gob.ar/contenido/revista/ediciones/47/cadenas/r47_06_TomateIndustrializado.pdf.

Ghadamgahi, F., Tarighi, S., Taheri, P., Saripella, G. V., Anzalone, A., Kalyandurg, P. B., Catara, V., Ortiz, R., & Vetukuri, R. R. (2022). Plant Growth-Promoting Activity of *Pseudomonas aeruginosa* FG106 and Its Ability to Act as a Biocontrol Agent against Potato, Tomato and Taro Pathogens. *Biology*, *11*(1), 140. https://doi.org/10.3390/biology11010140.

Harman, G. E., Howell, C. R., Viterbo, A., Chet, I., & Lorito, M. (2004). *Trichoderma* species - Opportunistic, avirulent plant symbionts. *Nature Reviews Microbiology* 2(1), 43–56.

Harman, G. E. (2006). Overview of mechanisms and uses of *Trichoderma* spp. Part I. Biological control and commercial applications. *European Journal of Plant Pathology 114*(1), 1-13. https://doi.org/10.1038/nrmicro797.

Hermosa, R., Belén Rubio, M., Cardoza, R. E., Nicolás, C., Monte, E., & Gutiérrez, S. (2013). The contribution of *Trichoderma* to balancing the costs of plant growth and defense. *International Microbiology* 16, 69–80. https://doi.org/10.2436/20.1501.01.181.

Hermosa, R., Viterbo, A., Chet, I., & Monte, E. (2012). Plant-beneficial effects of *Trichoderma* and of its genes. *Microbiology* 158, 17–25. https://doi.org/10.1099/mic.0.052274- 0.

Höfte, M. (2021). The use of *Pseudomonas* spp. As bacterial biocontrol agents to control plant diseases. In Wageningen University & Research, The Netherlands & J. Köhl (Eds.), *Burleigh Dodds Series in Agricultural Science* (pp. 301–374). Burleigh Dodds Science Publishing. https://doi.org/10.19103/AS.2021.0093.11.

Howell, C. R. (2003). Mechanisms employed by *Trichoderma* species in the biological control of plant diseases: the history and evolution of current concepts. Plant Disease *87*(1), 4-10.

Hussein, M. A. M., Abdel-Aal, A. M. K., Rawa, M. J., Mousa, M. A. A., Moustafa, Y. M. M., & Abo-Elyousr, K. A. M. (2023). Enhancing chili pepper (*Capsicum annuum* L.) resistance and yield against powdery mildew (*Leveillula taurica*) with beneficial bacteria. *Egyptian Journal of Biological Pest Control*, *33*(1), 114. https://doi.org/10.1186/s41938-023-00758-0.

Hyder, S., Gondal, A. S., Rizvi, Z. F., Atiq, R., Haider, M. I. S., Fatima, N., & Inam-ul-Haq, M. (2021). Biological Control of Chili Damping-Off Disease, Caused by *Pythium myriotylum*. *Frontiers in Microbiology*, *12*. https://doi.org/10.3389/fmicb.2021.587431.

Jackson, S. L., Whipps, J. M., & Lynch, J. M. (1991). Effects of formulation on the survival and biocontrol activity of *Trichoderma harzianum*. *Biocontrol Science and Technology*, *1*(2), 107-114.

Jena, N. (2012). *Pseudomonas* and *Trichoderma*: The most effective biocontrol agents against damping off pathogens of vegetable crops. *Biotechnology and Environmental Sciences*, *14*(2), 295–298.

Jeyarajan, R. (2006). Mass production of *Trichoderma*. In *Biological control of plant diseases* (pp. 167-173). Springer, Dordrecht.

Jia, Y., Niu, H., Zhao, P., Li, X., Yan, F., Wang, C., & Qiu, Z. (2023). Synergistic biocontrol of *Bacillus subtilis* and *Pseudomonas fluorescens* against early blight disease in tomato. *Applied Microbiology and Biotechnology*, *107*(19), 6071–6083. https://doi.org/10.1007/s00253-023-12642-w.

Jiménez-Pérez, O., Gallegos-Morales, G., Hernández-Castillo, F. D., Espinoza-Ahumada, C. A., Castro del Angel, E., Sánchez-Yañez, J. M., Jiménez-Pérez, O., Gallegos-Morales, G., Hernández-Castillo, F. D., Espinoza-Ahumada, C. A., Castro del Angel, E., & Sánchez-Yañez, J. M. (2023). Antagonistic activity of *Pseudomonas donghuensis* and *Bacillus subtilis* for the management of “Damping off” phytopathogens of the chile crop. *Revista BioCiencias*, *10*. https://doi.org/10.15741/revbio.10.e1382.

Kovács, C., Csótó, A., Pál, K., Nagy, A., Fekete, E., Karaffa, L., Kubicek, C. P., & Sándor, E. (2021). The biocontrol potential of endophytic *Trichoderma* fungi isolated from Hungarian grapevines. Part I. isolation, identification and *in vitro* studies. *Pathogens* 10,1612. doi: 10.3390/pathogens10121612.

Kulimushi, S. M., Muiru, W. M., & Mutitu, E. W. (2021). Potential of *Trichoderma* spp., *Bacillus subtilis* and *Pseudomonas fluorescens* in the management of early blight in tomato. *Biocontrol Science and Technology*, *31*(9), 912–923. https://doi.org/10.1080/09583157.2021.1900784.

Lee, S., & Lee, J. (2015). Beneficial bacteria and fungi in hydroponic systems: Types and characteristics of hydroponic food production methods. *Scientia Horticulturae*, *195*, 206–215. https://doi.org/10.1016/j.scienta.2015.09.011.

Lee, S. W., Ahn, I. P., Sim, S. Y., Lee, S. Y., Seo, M. W., Kim, S., Park, S. Y., Lee, Y. H., & Kang, S. (2010). *Pseudomonas* sp. LSW25R, antagonistic to plant pathogens, promoted plant growth, and reduced blossom-end rot of tomato fruits in a hydroponic system. *European Journal of Plant Pathology*, *126*(1), 1–11. https://doi.org/10.1007/s10658-009-9514-3.

Lenucci, M. S., Cadinu, D., Taurino, M., Piro, G., & Dalessandro. G. (2006). Antioxidant composition in cherry and high-pigment tomato cultivars. *Journal of Agricultural and Food Chemistry* 54,2606-2613.

Li, C., Gao, X., Huo, Y., Asseri, T. A. Y., Tian, X., & Luo, K. (2024). Evaluation of biocontrol efficacy of rhizosphere *Pseudomonas aeruginosa* for management of *Phytophthora capsici* of pepper. *PloS One 19*(9), e0309705. https://doi.org/10.1371/journal.pone.0309705.

Li, F., Ghanizadeh, H., Song, W., Miao, S., Wang, H., Chen, X., Liu, J., & Wang, A. (2023). Combined use of *Trichoderma harzianum* and *Clonostachys rosea* to manage Botrytis cinerea infection in tomato plants. *European Journal of Plant Pathology*, *167*(4), 637–650. https://doi.org/10.1007/s10658-023-02732-w.

Loayza, I. (2001). *Capsicum* y sus derivados en Iberoamérica: aspectos agrícolas, científicos, tecnológicos y económicos. CONACYT-Bolivia; CYTED, Subprograma IV Biomasa como Fuente de Productos Químicos y Energía.

Lommen, W. J. M. (2007). The canon of potato science: 27. Hydroponics. *Potato Research*, *50*(3–4), 315–318. https://doi.org/10.1007/s11540-008-9053-x.

Lopes, F. A. C., Steindorff, A. S., Geraldine, A. M., Brandao, R. S., Monteiro, V. N., Júnior, M. L., Coelho, A. S. G., Ulhoa, C. J. and Silva, R. N. (2012). Biochemical and metabolic profiles of *Trichoderma* strains isolated from common bean crops in the Brazilian Cerrado, and potential antagonism against *Sclerotinia sclerotiorum*. *Fungal Biology 116*(7), 815–824. https://doi.org/10.1016/j.funbio.2012.04.015.

López, M., Mazzola, M., & Mulero-Aparicio, A. (2010). Biocontrol mechanisms of *Trichoderma* against *Fusarium* and *Pythium* in hydroponic crops. *Journal of Plant Pathology*, *92*(1), 85–95.

Maboko, M., Du Plooy, C., & Chiloane, S. (2012). Effect of plant population, stem and flower pruning on hydroponically grown sweet pepper in a shade net structure. *African Journal of Agricultural Research* 7, 1742–1748. https://doi.org/10.5897/AJAR11.2078.

Mahapatra, B., Ghosh, D., & Mukhopadhyay, R. (2024). Biocontrol potential of *Trichoderma* and *Pseudomonas*: A review. *Biocontrol Potential of Trichoderma and Pseudomonas: A Review International Journal of Advanced Biochemistry Research*, *4*, 129–132.

Maruyama, C. R., Bilesky-José, N., de Lima, R. & Fraceto, L. F. (2020). Encapsulation of *Trichoderma harzianum* preserves enzymatic activity and enhances the potential for biological control. *Frontiers in Bioengineering and Biotechnology* 8, 225. doi: 10.3389/fbioe.2020.00225.

Mekonnen, H., Kibret, M. & Assefa, F. (2022). Plant Growth Promoting Rhizobacteria for Biocontrol of Tomato Bacterial Wilt Caused by *Ralstonia solanacearum*. *International Journal of Agronomy*, *2022*(1), 1489637. https://doi.org/10.1155/2022/1489637.

Molitor, H. (1990). The European perspective with emphasis on subirrigation and recirculation of water and nutrients. *Acta Horticulturae*, *272*, 165–174. https://doi.org/10.17660/actahortic.1990.272.24.

Muñoz-Celaya, A. L., Ortiz-García, M., Vernon-Carter, E. J., Jauregui-Rincón, J., Galindo, E. & Serrano-Carreón, L. (2012). Spray-drying microencapsulation of *Trichoderma harzianum* conidia in carbohydrate polymer matrices. Carbohydrate Polymers 88, 1141-1148.

Naher, L., Yusuf, U. K., Ismail, A., and Hossain, K. (2014). *Trichoderma* spp.: a biocontrol agent for sustainable management of plant diseases. *Pakistan Journal of Botany 46*(4), 1489-1493.

Orus, A. (2023). Ranking de las hortalizas más producidas a nivel mundial en 2021. https://es.statista.com/estadisticas/1308589/ranking-de-las-hortalizas-mas-producidas-a-nivel-mundial/.

Pandit, M. A., Kumar, J., Gulati, S., Bhandari, N., Mehta, P., Katyal, R., Rawat, C. D., Mishra, V., & Kaur, J. (2022). Major Biological Control Strategies for Plant Pathogens. *Pathogens*, *11*(2), 273. https://doi.org/10.3390/pathogens11020273.

Panthee, D. R. and Chen, F. (2010). Genomics of fungal disease resistance in tomato. *Current Genomics* 11, 30–39.

Pastor, N., Carlier, E., Andrés, J., Rosas, S. B., & Rovera, M. (2012). Characterization of rhizosphere bacteria for control of phytopathogenic fungi of tomato. *Journal of Environmental Management*, *95 Suppl*, S332-7. https://doi.org/10.1016/j.jenvman.2011.03.037.

Patel, J. S., Yadav, S. K., Bajpai, R., Teli, B., & Rashid, M. (2020). Pgpr secondary metabolites: an active syrup for improvement of plant health. Molecular aspects of plant beneficial microbes in agriculture. *Elsevier*, *195–208*.

Peralta, I. E. and Spooner, D. M. (2007). History, origin and early cultivation of tomato (*Solanaceae*). In: Genetic Improvement of Solanaceous Crops. Science Publishers. Enfield, New Hampshire, USA. Vol. 2, 1–24.

Pothiraj, G., Hussain, Z., Singh, A. K., Solanke, A. U., Aggarwal, R., Ramesh, R., & Shanmugam, V. (2021). Characterization of Fusarium Spp. Inciting Vascular Wilt of Tomato and Its Management by a Chaetomium-Based Biocontrol Consortium. *Frontiers in Plant Science*, *12*. https://doi.org/10.3389/fpls.2021.748013.

Raaijmakers, J. M., Paulitz, T. C., Steinberg, C., Alabouvette, C., & Moënne-Loccoz, Y. (2009). The rhizosphere: a playground and battlefield for soilborne pathogens and beneficial microorganisms. *Plant and Soil*, *321*(1–2), 341–361. https://doi.org/10.1007/s11104-008-9568-6.

Ramanujam, B., Prasad, R. D., Sriram. S. & Rangeswaran, R. (2010). Mass production, formulation, quality control and delivery of *Trichoderma* for plant disease management. *The Journal of Plant Protection Science 2*(2), 1-8.

Rana, A., Kumar, N., Roy, R., Kumari, A., Kumar, R., Gautam, A., & Das, S. (2025). Field efficacy of *Trichoderma viride*, *Pseudomonas fluorescens*, and consortia of microbial agents against tomato late blight, *Phytophthora infestans* (Mont.) de Bary in Indo-Gangetic Plain of Bihar. *Crop Protection (Guildford, Surrey)*, *190*(107128), 107128. https://doi.org/10.1016/j.cropro.2025.107128.

Renault, D., Déniel, F., Benizri, E., Sohier, D., Barbier, G., & Rey, P. (2007). Characterization of *Bacillus* and *Pseudomonas* strains with suppressive traits isolated from tomato hydroponic-slow filtration unit. *Canadian Journal of Microbiology*, *53*(6), 784–797. https://doi.org/10.1139/W07-046.

Rini, C. R., & Sulochana, K. (2006). Management of seedling rot of chilli (Capsicum annuum L.) using Trichoderma spp. and fluorescent pseudomonads (Pseudomonas fluorescens). *Journal of Tropical Agriculture*, *44*(0), 79–82. https://jtropag.kau.in/index.php/ojs2/article/view/159.

Rokni, N., Shams Alizadeh, H., Bazgir, E., Darvishnia, M., & Mirzaei-Najafgholi, H. (2021). The tripartite consortium of *Serendipita indica*, *Trichoderma simmonsii,* and bell pepper (*Capsicum annum*). *Biological Control*, *158*, 104608. https://doi.org/10.1016/j.biocontrol.2021.104608.

Rovera, M., Pastor, N., Niederhauser, M. & Rosas, S. B. (2014). Evaluation of *Pseudomonas chlororaphis* subsp. *aurantiaca* SR1 for growth promotion of soybean and for control of *Macrophomina phaseolina*. *Biocontrol Science and Technology*, *24*(9), 1012–1025. https://doi.org/10.1080/09583157.2014.910293.

Saravanakumar, K., Fan, L., Fu, K., Yu, C., Wang, M., Xia, H,, Sun, J., Li, Y., & Chen, J. (2016). Cellulase from *Trichoderma harzianum* interacts with roots and triggers induced systemic resistance to foliar disease in maize. *Scientific Reports* 6,35543. doi: 10.1038/srep35543.

Sharma, B. L., Singh, S. P. and Sharma, M. L. (2012). Bio-degradation of crop residues by *Trichoderma* species vis-a-vis nutrient quality of the prepared compost. *Sugar Technology 14*(2), 174-180.

Shoresh, M., Harman, G. E., & Mastouri, F. (2010). Induced systemic resistance and plant responses to fungal biocontrol agents. *Annual Review of Phytopathology*, *48*(1), 21–43. https://doi.org/10.1146/annurev-phyto-073009-114450.

Singh, V. K., Singh, A. K., & Kumar, A. (2017). Disease management of tomato through PGPB: current trends and future perspective. *3 Biotechnology* *7*(4). https://doi.org/10.1007/s13205-017-0896-1.

Sood, D., Sharma, A., & Sharma, M. (2021). *Management of bacterial wilt of tomato through a consortium of biocontrol agents*. *Journal of Pharmacognocy and Phytochemistry* 10(1):1355–1358.

Suresh, P., Shanmugaiah, V., Rajagopal, R., Muthusamy, K. & Ramamoorthy, V. (2022). *Pseudomonas fluorescens* VSMKU3054 mediated induced systemic resistance in tomato against *Ralstonia solanacearum. Physiological and Molecular Plant Pathology*, *119*, 101836. https://doi.org/10.1016/j.pmpp.2022.101836.

Suresh, P., Vellasamy, S., Almaary, K. S., Dawoud, T. M. & Elbadawi, Y. B. (2021). *Fluorescent pseudomonads* (FPs) as a potential biocontrol and plant growth promoting agent associated with the tomato rhizosphere. *Journal of King Saud University - Science*, *33*(4), 101423. https://doi.org/10.1016/j.jksus.2021.101423.

Taha, N. A., Elsharkawy, M. M., Shoughy, A. A., El-Kazzaz, M. K., & Khedr, A. A. (2023). Biological control of postharvest tomato fruit rots using *Bacillus* spp. And *Pseudomonas* spp. *Egyptian Journal of Biological Pest Control*, *33*(1), 106. https://doi.org/10.1186/s41938-023-00752-6.

Taylor, A. G., Harman, G. E. & Nielsen, P. A. (1994). Biological seed treatments using *Trichoderma harzianum* for horticultural crops. *HortTechnology* *4*(2), 105-109.

Vinale, F., Sivasithamparam, K., Ghisalberti, E. L., Marra, R., Woo, S. L., & Lorito, M. (2008). Trichoderma–plant–pathogen interactions. *Soil Biology & Biochemistry*, *40*(1), 1–10. https://doi.org/10.1016/j.soilbio.2007.07.002.

Wang, X., Zhou, X., Cai, Z., Guo, L., Chen, X., Chen, X., Liu, J., Feng, M., Qiu, Y., Zhang, Y., & Wang, A. (2020). A Biocontrol Strain of Pseudomonas aeruginosa CQ-40 Promotes Growth and Controls Botrytis cinerea in Tomato. *Pathogens*, *10*(1), 22. https://doi.org/10.3390/pathogens10010022.

Whipps, J. M. (2001). Microbial interactions and biocontrol in the rhizosphere. *Journal of Experimental Botany* *52*(1), 487-511.

Woo, S. L., Scala, F., Ruocco, M. & Lorito, M. (2006) The molecular biology of the interactions between *Trichoderma* spp., phytopathogenic fungi, and plants. *Phytopathology* 96, 181–185.

Yadav, M., Dubey, M. K., & Upadhyay, R. S. (2021). Systemic Resistance in Chilli Pepper against Anthracnose (Caused by *Colletotrichum truncatum*) Induced by *Trichoderma harzianum*, *Trichoderma asperellum* and *Paenibacillus dendritiformis*. *Journal of Fungi (Basel, Switzerland)*, *7*(4), 307. https://doi.org/10.3390/jof7040307.

Yedidia, I., Srivastva, A. K., Kapulnik, Y., & Chet, I. (2001). Effect of *Trichoderma harzianum* on microelement concentrations and increased growth of cucumber plants. *Plant Soil* 235:235–242.

Yendyo, S., Ramesh, & Pandey, B. R. (2018). Evaluation of *Trichoderma* spp., Pseudomonas fluorescens, and Bacillus subtilis for biological control of Ralstonia wilt of tomato. *Food Research* 6: 2028. https://doi.org/10.12688/f1000research.12448.3.

(N.d.). Morningagclips.com. Retrieved May 26th, 2025, from https://www.morningagclips.com/vmr-hydroponics-market-to-hit-usd-58-3-billion-by-2030/#:~:text=According%20to%20a%20report%20by,7.5%25%20between%202023%20and%202030.

Chapter 6

In Vitro Antagonistic Activity of Microorganisms Isolated from Sorghum on Phytopathogenic Strains of *Epicoccum sorghinum*

Gortari María Cecilia[1,2]
Vio Santiago Adolfo[1,3]
Luna María Flavia[1,2]
and Astoreca Andrea[3,*]

[1]Centro de Investigación y Desarrollo en Fermentaciones Industriales (CINDEFI), Universidad Nacional de La Plata (UNLP), La Plata, Buenos Aires, Argentina
[2]Comisión de Investigaciones Científicas (CIC-PBA), La Plata, Buenos Aires, Argentina
[3]Consejo Nacional de Investigaciones Científicas y Técnicas (CONICET), Ciudad Autónoma de Buenos Aires (CABA), Argentina

Abstract

Sorghum is a cereal of global importance that plays a fundamental role in the Argentine agro-industrial chain. Fungal species belonging to the *Epicoccum* genus that produce tenuazonic acid (TeA) have been reported in this crop, with the consequent agroeconomic impact. An alternative to such contamination in this crop is the use of bioinputs that act as fungal biocontrol agents. The objective of this work was to isolate and evaluate the antagonistic potential of twenty-five bacterial isolates and four yeast isolates against four *E. sorghinum* isolates. All microorganisms were isolated from sorghum grains and molecularly identified. The capacity

---

* Corresponding Author's Email: astoreca@biotec.quimica.unlp.edu.ar

In: Biocontrol Agents (BCAs)
Editor: Daiana García
ISBN: 979-8-90134-102-5

for growth control of the pathogen was evaluated by confronting it with four *E. sorghinum* isolates in plate culture. The test was considered positive when the microorganism stopped fungal growth. With those that showed activity, a dual culture was performed, and the percentage of inhibition of radial growth (PIRG) was calculated. On the other hand, their capacity to produce lytic enzymes and volatile compounds was evaluated. Enzyme production was carried out in a solid minimal medium with the addition of inducing substrates to detect chitinolytic, amylolytic, proteolytic, and lipolytic activity. The detection of chitinase, amylases, and proteases was evidenced by the formation of a transparency halo, while lipolytic activity was detected by the presence of an opacity halo due to the precipitation of calcium salts present in the medium. The changes in the macromorphological characteristics of fungal colonies were observed with the use of a stereoscopic microscope. For the production of volatile compounds, each fungus was confronted with each of the microorganisms in devices specially designed for this purpose. Of the total number of microorganisms tested, three bacterial isolates presented a PIRG indicative of antagonistic activity. Regarding enzymatic activity, only one bacterial isolate was positive for the four enzymes tested. The production of volatile compounds was manifested with macro and micro morphological changes in both microbial groups studied, without showing obvious alterations in the *E. sorghinum* growth. These results reflect a state of stress of the pathogen against the microorganisms. Although it is not sufficient to consider any of them as antagonistic, their bibliographical reference as such merits the deepening and expansion of the studies carried out on the activity of these potential agents on TeA production.

Keywords: Sorghum, *Epicoccum*-tenuazonic acid, bioinputs, biocontrol agents

Introduction

Sorghum (*Sorghum bicolor* L. *Moench*) is recognized for its high agronomic productivity and its nutritional and bioactive properties, superior to those of cereals such as rice, wheat, and corn (Aruna et al., 2018). Sorghum is cultivated in more than 100 countries worldwide due to its resilience in dry and arid environments and its adaptability to diverse climates. For this reason, sorghum is the best option to guarantee a sustainable food system with minimal resource use, reducing the impact on the environment and providing food and nutritional security (Davis et al., 2018). Globally, the United States,

Australia, and Argentina are the main sorghum exporters, representing 59.2%, 23.3%, and 12.1% of total exports, respectively, while China, Japan, and Spain are importers (Rohlupuii, 2024). However, in the last decade, the area cultivated with sorghum has decreased drastically in many major producing countries due to a deterioration in grain quality caused by the presence of fungi (Khaskheli et al., 2025), which also have the ability to produce mycotoxins, making it impossible to use the grain for animal feed (https://www.fao.org/4/x5027s/x5027S03.htm). A promising strategy to solve this problem, which is being promoted all over the world, not only to prevent the action of pathogenic fungi on grains but also on plants and fruits, is the use of bio-inputs. Bio-inputs based on microorganisms such as bacteria and fungi are an environmentally friendly, low-cost, and sustainable alternative to hazardous chemical inputs (Denning, 2025). It is possible to categorize them as biocontrol agents, biofertilizers, and biostimulants according to their mode of action. In particular, the bioprospection of microorganisms with potential to control plant diseases and grain contamination has become a central point of agricultural research (Dornelas et al., 2023). Biocontrol microorganisms (BCM) can prevent the growth of fungi and/or counteracting the effect of mycotoxins, thereby contributing to the agricultural sector, animal production, and public health in general, which are indirectly affected by the presence of fungi and mycotoxins in numerous crops. BCM can improve the phytosanitary status of grain and plants, inhibit the development of pathogens through spatial and/or nutritional competition, parasitism, production and release of antibiotics, hydrolytic enzymes, volatile compounds, and biofilm formation (Ramirez-Pool et al., 2024). In recent years, tenuazonic acid (TeA), a little-studied mycotoxin, has been associated with the contamination of sorghum grains and derivatives (Asam and Rychlik, 2013; Oliveira et al., 2017; Mohammed et al., 2022). Its presence was recently reported in 100% of the analyzed sorghum samples (Hipperdinger et al., 2024). Currently, it has been observed that *Epicoccum sorghinum* is an important producer of TeA and is also considered one of the main components of the complex of fungal diseases that affect sorghum grain (Oliveira et al., 2017). In this context, the use of BCM isolated from sorghum grains might be a viable, promising, and ecological alternative for controlling plant diseases produced by *E. sorghinum*.

The aim of this work was to isolate and characterize bacteria and yeasts associated with sorghum grains, focusing on the *in vitro* antagonistic potential and mode of interaction with *E. sorghinum* as potential biotechnological alternatives for controlling their growth in sorghum grains.

Material and Methods

Sample Collection

Seven samples of red, brown, and white sorghum grain samples were collected from the Experimental Station of the National Institute of Agricultural Technology (INTA) in Manfredi, Córdoba, Argentina, to isolate bacteria and yeast. In the laboratory, the grains were finely ground in a Buehler mill for microbiological analysis.

Microorganism Isolation and Conservation

Bacteria and yeast isolation was performed on solid culture media using the surface-spread method by shaking 10 g portions of each grounded sorghum grain sample with 90 mL of 0.1% peptone water solution for 30 minutes. Serial dilutions of 10^{-1} to 10^{-4} concentration were made in 9 mL of 0.1% peptone water solution, and 0.1 mL aliquots were inoculated in triplicate onto petri dishes with Dichlorinated Nutrient agar, and YPD agar (yeast extract 1%, peptone 2% and dextrose 2%) with 100 mgL^{-1} of chloramphenicol to isolate bacteria and yeast isolation, respectively. Plates were incubated at 35°C for 48 hours for bacteria isolation and at 25°C for seven days for yeast isolation, both in darkness. Once incubation time was over, different colonies were selected based on their morphological characteristics -color, shape, size, regular or irregular, and convex or flat (Li et al., 2021), and re-isolated until axenic bacterial and yeast isolates were obtained. Isolates were stored at -80°C with 20% glycerol.

Taxonomic Assignment of Bacterial and Yeast Isolates

Thermal cell lysates from bacterial isolates were used to partially amplify the 16S rDNA gene by PCR, using 27F (3'-AGAGTTTGATCMTGGCTCAG-5') and 907R (3'-CCGTCAATTCMTTTRAGTTT-5') primers. Amplicons were sequenced at Macrogen Inc. (Seoul, Korea), and the obtained 16S rDNA sequences were aligned and assembled using BioEdit v 7.0.5.3 software and then analyzed using the BlastN tool in the NCBI database for taxonomic assignment. Yeast genomic DNA was extracted according to the method described by Schena et al. (1999) with some modifications. Two amplicons of the divergent D1/D2 region of yeast 26S rDNA were amplified by PCR using NL1 (5'- GCATATCAATAAGCGGAGGAAAAG-3'), NL4 (5'-GGTCCGTGTTTCAAGACGG-3'), ITS1 (5′-TCCGTAGGTGAACC TGCGG-3′), and ITS4 (5′-TCCTCCGCTTATTGATATGC-3′) primers (Kurtzman and Robnett, 1998). Sequencing and subsequent analysis were the same as described above for bacterial isolates.

In Vitro Characterization of Isolates' Antagonistic Activity

Determination of Antagonism-Related Enzymatic Activities

Chitinase, protease, amylase, and lipase activities of bacterial and yeast isolates were evaluated on Petri plates with solid minimal medium with the addition of the corresponding substrates (chitin, nonfat milk powder, starch, and tween 80, respectively) according to the methodology described by Vio (2023). Detection of chitinolytic, proteolytic, and amylolytic activity was evidenced by the formation of a transparent zone surrounding the microbial colony, while lipolytic activity was indicated by the presence of an opaque zone surrounding the microbial colony due to the precipitation of calcium salts. Changes in the culture media were detected by direct observation and confirmed using a stereomicroscope. Bacterial or yeast isolates produce hydrolytic enzymes that were qualitatively categorized as positive (+) or negative (-).

Qualitative Fungal Growth Inhibition Screening

To qualitatively evaluate the ability of bacterial and yeast isolates to inhibit the growth of four phytopathogenic strains of *E. sorghinum*, an *in vitro* assay was performed on potato dextrose agar (PDA, Oxoid) plates. Four phytopathogenic fungal strains of *E. sorghinum* of the Collection of the Mycotoxins Laboratory of the Center for Research and Development in Industrial Fermentations (CINDEFI-CONICET-UNLP) (Hipperdinger et al., 2024) were grown on PDA plates for 7 days at 28°C, while bacterial and yeast isolates were grown for 24 hours at 30°C on Nutrient Agar and YPD agar plates, respectively. An 8 mm diameter mycelium-agar disc of each *E. sorghinum* strain was placed in the center of a Petri plate amended with PDA medium, and bacterial and yeast isolates were streaked perpendicularly on opposite sides at a distance of 3 cm from the pathogen (Figure 1a). Plates were incubated at 30°C for 5-7 days. Control plates were prepared by growing each tested fungus in the absence of bacteria or yeast. Bacterial and yeast isolates were characterized by positive (+) or negative (-) based on a qualitative mycelial growth inhibition.

Antagonistic Activity Quantification by Dual Culture Plate Assay

Microbial isolates characterized as positive (+) in the qualitative screening of *E. sorghinum* growth inhibition were selected to quantify this antagonistic activity by a dual culture plate assay. Selected isolates alone faced with each pathogenic fungus on PDA plates, spacing 5 cm between an 8 mm diameter mycelium-agar disc of each *E. sorghinum* strain and a streak of each microbial isolate (Figure 1b). Plates were incubated at 30°C for 5-7 days. Control plates were prepared by growing each tested fungus in the absence of bacteria or yeast. The percentage of inhibition of radial growth (PIRG) was calculated in duplicate as follows (Anith et al., 2021):

$$PIRG\ (\%) = [(R_c - R_i)/(R_c)] \times 100$$

Where:

R_c = Radial distance of the fungal colony on the control plate

R_i = Radial distance of the fungal colony on each dual culture plate with bacteria or yeast isolate

Fungal Growth Inhibition by Volatile Organic Compounds (VOCs)

The production of antagonistic VOCs by bacterial and yeast isolates was evaluated *in vitro* assays. For this purpose, a device consisting of two plates of equal diameter, each containing PDA culture medium, was designed. An 8 mm diameter mycelium-agar disc of each *E. sorghinum* strain was placed in the center of one plate, and the different microbial isolates were streaked on the other plate. Finally, the device was closed by placing one plate over the other, leaving the gaseous phase connecting the pathogen with the microbial isolation (Figure 1c). Plates were incubated at 30°C for 5-7 days. Control plates were prepared by growing each tested fungus in the absence of bacteria or yeast, and the colony diameter of the pathogenic fungus was measured at the end of the assay to calculate the PIRG in duplicate.

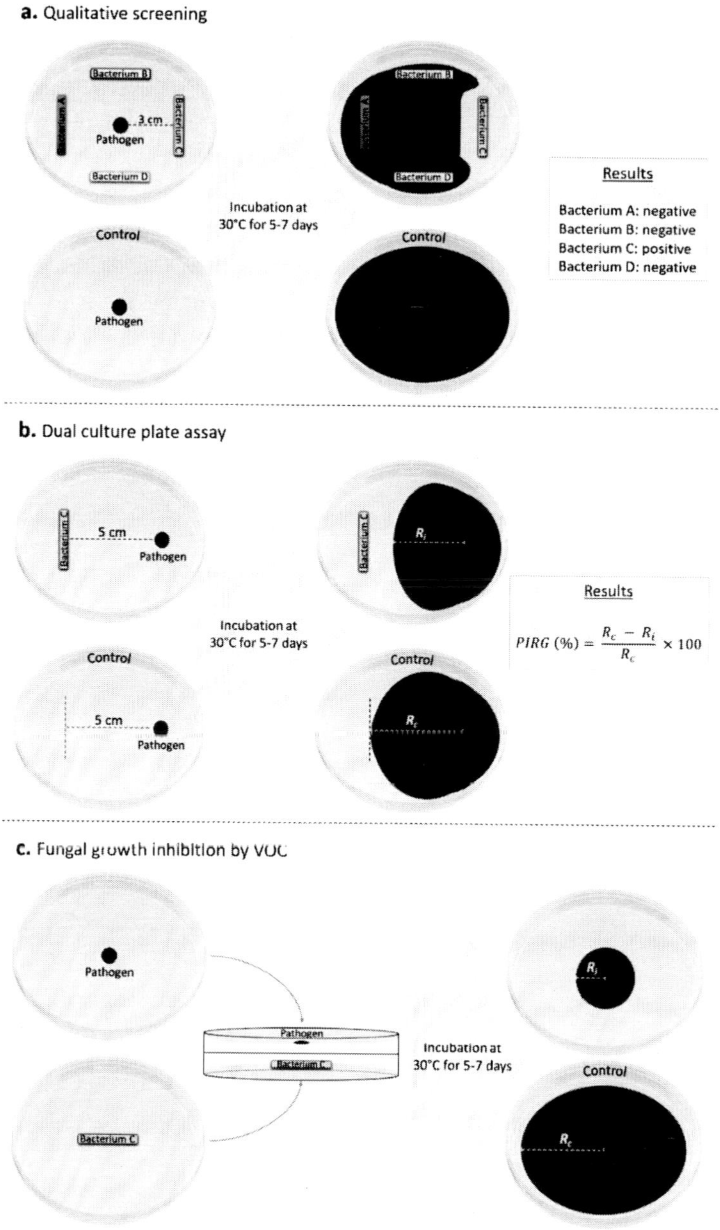

Figure 1. Diagram of *in vitro* bacterial and yeast antagonism assays against different strains of *Epicoccum sorghinum*. (a) Initial qualitative screening; (b) Dual confrontation and quantification of percentage radial growth inhibition (PIRG); (c) Fungal growth inhibition by volatile organic compounds (VOCs).

Results and Discussion

The dual challenge of mitigating substantial crop losses caused by phytopathogenic microorganisms and reducing the reliance on synthetic pesticides, with their associated environmental and health risks, has intensified the search for sustainable agricultural practices. In this context, biological control, using natural antagonistic microorganisms, for example, through bioinputs, represents a promising avenue. Currently, research efforts are directed toward developing these agents, based on microorganisms isolated from the environments where the pathogen to be controlled is found, since these agents would perform better than foreign/exotic microorganisms (Ayaz et al., 2023; Ezzati Yazid et al., 2023). In this context, a total of 29 microorganisms were isolated from different samples of brown, red, and white sorghum grains, including 24 bacterial isolates belonging to the genera *Microbacterium*, *Pantoea*, *Curtobacterium*, *Pseudomonas*, *Bacillus*, *Enterobacter*, and *Kosakonia*, and four yeast isolates identified as *Hannaella luteola*, *Papiliotrema flavescens*, and *Meyerozyma caribbica* (Table 1). All microorganisms were deposited in the Collection of the Mycotoxins Laboratory of the Center for Research and Development in Industrial Fermentations (CINDEFI-CONICET-UNLP) (Hipperdinger et al., 2024).

Table 1. Results of taxonomic assignment and screening assays related to the antagonistic activity of 29 microbial isolates from different sorghum grain samples

Isolate	Sample	Taxonomic assignment			Qualitative (-) and dual culture plate assay (PIRG)				Fungal growth inhibition by VOC (PIRG)			
		Genus	Identity (%)	Accession number	LMCIN-18.4	LMCIN-15.2	LMCIN-9.4	LMCIN-3.11	LMCIN-18.4	LMCIN-15.2	LMCIN-9.4	LMCIN-3.11
LMCIN-N1		*Microbacterium* sp.	99,9	MT007290.1	-	-	-	-	NI	9±2	NI	52±8
LMCIN-A2		*Curtobacterium* sp.	99,5	CP041259.1	-	-	-	-	17±2	24±2	16±6	5±2
LMCIN-A3	M228 Brown grain	*Curtobacterium* sp.	100,0	CP041259.1	-	-	-	-	2±1	NI	57±2	16±2
LMCIN-B1		*Enterobacter* sp.	99,9	AP022628.1	-	-	-	-	4±1	NI	10±3	10±3
LMCIN-AAR		*Meyerozyma caribbica*	100,0	MT251155.1	26±2	15±3	14±5	6±2	NI	NI	NI	NI
LMCIN-X9		*Pantoea* sp.	99,2	MZ501277.1	10±1	-	-	-	11±1	25±7	13±2	7±2
LMCIN-AIA		*Pseudomonas* sp.	99,5	MH362752.1	-	-	-	-	NI	14±1	18±6	17±3
LMCIN-X8		*Kosakonia* sp.	99,9	MK649841.1	-	-	-	11±1	2±1	47±6	21±5	10±4
LMCIN-X11	M283 Brown grain	*Pantoea* sp.	99,8	MN636697.1	14±4	17±1	-	-	5±2	14±2	37±6	15±1
LMCIN-X10		*Bacillus* sp.	100,0	MN826587.1	-	-	-	-	NI	46±5	27±3	14±5
LMCIN-X7		*Kosakonia* sp.	99,8	CP035129.1	-	-	57±4	14±2	6±2	4±2	14±1	7±3
LMCIN-L6		*Papiliotrema flavescens*	100,0	MT447858.1	-	-	-	-	6±3	9±1	NI	NI
LMCIN-L3		*Curtobacterium* sp.	99,7	OR039161.1	-	-	-	-	NI	NI	6±2	NI
LMCIN-X1	M273	*Pantoea* sp.	99,9	KU145674.1	10±4	12±3	-	-	9±1	26±3	50±4	25±5
LMCIN-L280	Brown grain	*Bacillus* sp.	100,0	CP012475.1	-	.-	-	-	11±4	24±6	22±6	21±2
LMCIN-L2		*Hannaella luteola*	100,0	MN124389.1	-	-	-	-	3±1	9±2	12±7	28±2
LMCIN-X4		*Pantoea* sp.	100,0	CP028033.1	-	-	-	-	37±5	19±2	20±2	18±4
LMCIN-X5		*Pantoea* sp.	99,9	MK398009.1	14±2	7±1	9±2	-	NI	15±5	19±5	9±3
LMCIN-X6	R269	*Pantoea* sp.	99,9	MN932344.1	10±4	-	-	-	28±4	5±3	22±2	9±4
LMCIN-X3	Red grain	*Curtobacterium* sp.	99,3	OR426296.1	-	17±6	9±2	11±3	22±3	11±6	48±8	9±4
LMCIN-L4		*Papiliotrema flavescens*	100,0	MT447858.1	-	-	-	-	35±2	9±2	NI	13±4
LMCIN-X2		*Pantoea* sp.	97,5	MH127816.1	12±2	19±2	-	-	7±2	NI	34±4	14±5
LMCIN-288b	R288	*Bacillus* sp.	100,0	CP051464.1	57±3	55±2	29±7	46±4	33±8	10±4	NI	13±5
LMCIN-288a	Red grain	*Bacillus* sp.	99,9	MN826587.1	-	-	-	-	21±4	16±5	31±2	15±7
LMCIN-R2		*Enterobacter* sp.	99,9	AP022628.1	10±2	-	9±4	-	26±5	4±2	37±2	13±2
LMCIN-280a	R280 Red grain	*Bacillus* sp.	93,0	OQ981424.1	-	-	-	-	33±2	27±2	27±1	36±3
LMCIN-280b		*Bacillus* sp.	100,0	LT906438.1	26±4	26±3	-	-	11±5	52±1	29±4	15±8
LMCIN-N2	B250	*Microbacterium* sp.	100,0	CP113067.1	-	-	-	-	NI	9±4	16±1	11±7
LMCIN-AAN	White grain	*Pantoea* sp.	99,0	MH127816.1	7±1	-	-	-	NI	22±3	16±5	75±9

This taxonomic diversity aligns with previous studies reporting a complex microbiota associated with sorghum grains, which includes both phytopathogenic and beneficial microorganisms (Wang et al., 2020; Bhise et al., 2021). Moreover, several genera are mentioned in various publications related to the research and development of more sustainable control alternatives for different fungal species and stages of several crops (Navarta et al., 2019; Ayaz et al., 2023; EL Alami and EL Attari, 2020; Taibi et al., 2020; Oztekin et al., 2023).

A first step in the search for organisms with potential biological control agents is screening for inhibitory activity against pathogens. The most common technique used is the dual culture assay on solid plate media. In this case, a slight modification of the technique was used to screen more than one bacterium on the phytopathogen to be controlled, as described in the work reported by Anith et al. (2021).

Following this initial screening, the corresponding dual cultures were performed. The resulting PIRG ranged from 0% to 57.2% under the tested conditions. Of the total bacteria tested, two isolates (LMCIN-X7 and LMCIN-288b) had an inhibition percentage equal to or greater than 30%. In the first case, *Kosakonia* sp. (PIRG = 57.2% on *E. sorghinum* LMCIN 9.4) is a bacterial genus recognized as plant growth-promoting bacteria (PGPB) with *in vitro* antifungal activity reported against phytopathogenic fungi such as *Botrytis cinerea*, *Penicillium expansum*, *Fusarium oxysporum*, *Colletotrichum gloeospoioides*, and *Pectobacterium aroidearum* (Lambrese et al., 2019; Taibi et al., 2020; Ali and El-baz, 2023; Mena Navarro et al., 2024). In the second case, *Bacillus* sp. showed PIRG between 28.6% and 57.2% on the 4 *E. sorghinum* tested isolates. It is a genus widely studied in relation to its potential ability to control the growth of phytopathogenic fungi and used globally as PGPB and control of pests and diseases caused by other pathogens (Ayaz et al., 2023; Dobrzynski et al., 2023; Mis et al., 2024; Vasques et al., 2024).

Regarding the yeasts analyzed, LMCIN-AAR (*Meyerozyma caribbica*) is a yeast genus that showed antagonism during the initial confrontations; however, this inhibition was not evident in the dual cultures. The *Meyerozyma* genus includes yeasts with antagonistic action on phytopathogenic and mycotoxigenic fungi (Ozteki et al., 2023).

Additional tests were performed on bacteria and yeast isolates: production of hydrolytic enzymes and volatile compounds to expand knowledge of the mechanisms involved in antagonism.

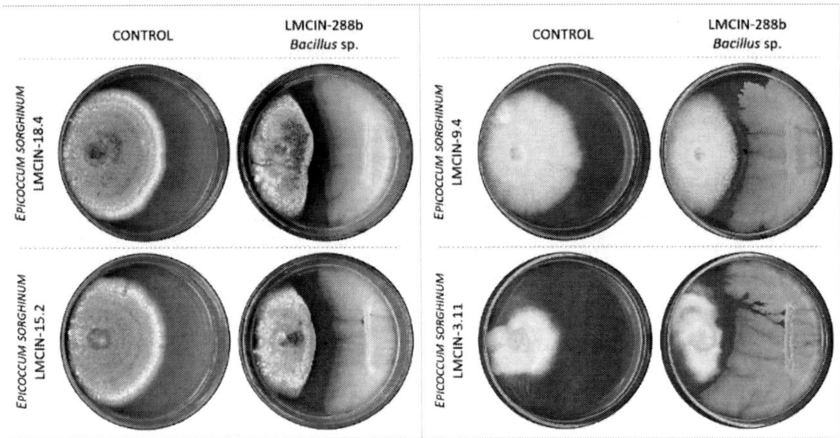

Figure 2. Photographs of the plates resulting from the dual culture assay between different strains of *E. sorghinum* and the isolate LMCIN-288b of the genus *Bacillus*.

Regarding enzymatic activity, the bacterium LMCIN-X7 (*Kosakonia* sp.) showed lipolytic activity, LMCIN-280b (*Bacillus* sp.) proteolytic activity, and only the LMCIN-288b strain (*Bacillus* sp.) demonstrated the broadest enzymatic activity, being positive for chitinase, protease, amylase, and lipase. In yeasts, although the production of hydrolytic enzymes is one of the main mechanisms of action in the control of post-harvest diseases (Leal Rodríguez et al., 2024), in this research, only lipolytic activity was recorded for LMCIN-L2 (*Hannaella luteola*), LMCIN-L4 (*Papiliotrema flavescens*), and LMCIN-L6 (*Papiliotrema flavescens*).

Production of volatile compounds by bacteria and yeasts was manifested, in some cases, by colony size reduction and phenotypic changes such as color, a different sporulation pattern, as well as high exudate production. The phenotypic changes produced by bacteria and yeasts, even without growth inhibition, likely reflect a stress state for the pathogen in the presence of these microorganisms, as previously reported by Vázquez-Gómez et al. (2019). Similar findings have been reported when weak antagonists or suboptimal VOC production conditions were present (Gotor-Vila et al., 2017).

Regarding colony size reduction, a wide range of inhibition percentages was obtained (1.16-7.5). The bacterial genus with the highest PIRG was *Pantoea* sp. This genus includes several species known as biocontrol agents, in which the production of volatile compounds with antimicrobial activity is one of the inhibition mechanisms (Duchateau et al., 2024). Another genus that proved promising in the control of phytopathogenic fungi due to its ability to produce a wide variety of volatile compounds was *Bacillus* sp. These results coincide with those obtained by Toral et al. (2021) who reported higher percentages of growth inhibition on different species belonging to *Alternaria*, *Botrytis*, *Fusarium*, *Monillinia* and *Sclerotinia*.

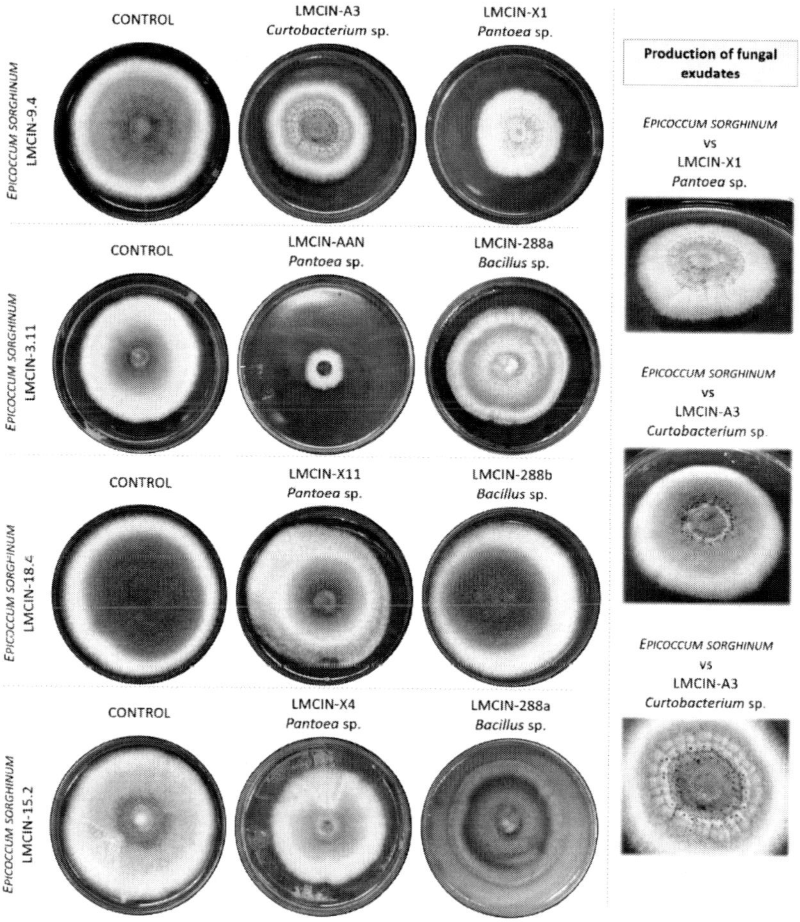

Figure 3. Photographs of the plates resulting from the VOC assay between different strains of *E. sorghinum* and different isolates that showed some growth inhibition effect or macromorphological change.

Since the large number of factors that modify the visualization of growth inhibition (media, seeding and culture conditions, evaluation methods, etc.), the production of hydrolytic enzymes (culture techniques and media) and the production of volatile compounds (physiological state of the microorganism, medium, and culture conditions), the results obtained constitute the beginning of the studies regarding the microorganisms studied.

Conclusion

This study provides a first insight into the considerable taxonomic diversity of microorganisms in the sorghum grain agroecosystem and highlights an emerging health problem caused by the presence of different *E. sorghinum* isolates and the consequent contamination with tenuazonic acid under storage conditions. Furthermore, the construction of a collection of microorganisms isolated from sorghum grains, including bacteria and yeast genera with known biotechnological potential, provides a very important basis for deepening and expanding our understanding of the mechanisms involved in the inhibition of *E. sorghinum* and the mycotoxin (TeA).

References

Ali, A., El-baz, G. A., (2023). Potential role of *Kosakonia radicincitans* to suppress *Fusarium oxysporum* in Maize plants. *Int. J. Chem. Biochem. Sci.* 24, 1434-1441.

Anith, K. N., Nysanth, N. S., Natarajan, C., (2021). Novel and rapid agar plate methods for *in vitro* assessment of bacterial biocontrol isolates' antagonism against multiple fungal phytopathogens. *Lett. App. Microbiol.* 73, 229-236.

Aruna, C., Das, I. K., Reddy, P. S., Ghorade, R. B., Gulhane, A. R., Kalpande, V. V., Kajjidoni, S. T., Hanamaratti, N. G., Chattannavar, S. N., Mehtre, S., Gholve, V., Kamble, K. R., Deepika, C., Kannababu, N., Bahadure, D. M., Govindaraj, M., Tonapi, V. A., (2021). Development of sorghum genotypes for improved yield and resistance to grain mold using population breeding approach. *Front. Plant Sci.* 28, 12:687332.

Aruna, C., Visarada, K. B. R. S., Venkatesh Bhat, B., Tonapi Vilas, A., (2018). Breeding sorghum for diverse end uses. Woodhead Publishing ISBN 978-0-08-101879-8.

Ayaz, M., Li, C.-H., Ali, Q., Zhao, W., Chi, Y.-K., Shafiq, M., Ali, F., Yu, X.-Y., Yu, Q., Zhao, J.-T., Yu, J.-W., Qi, R.-D., Huang, W.-K., (2023). Bacterial and Fungal Biocontrol Agents for Plant Disease Protection: Journey from Lab to Field, Current Status, Challenges, and Global Perspectives. *Molecules*, 28, 6735.

Bhise, P., Jadhav, P., Kale, R. (2021). Diversity and biocontrol potential of endophytic bacteria from sorghum (Sorghum bicolor L.). *Journal of Applied Microbiology*, 130(3), 923–937.

Davis, K. F., Chiarelli, D. D., Rulli, M. C., Chhatre, A., Richter, B., Singh, D., DeFries, R., (2018). Alternative cereals can improve water use and nutrient supply in India. *Sci. Advances* 4, eaao1108.

Denning, G., (2025). Sustainable intensification of agriculture: the foundation for universal food security. *Npj Sustainable Agric.* 3, 7.

Dobrzynski, J., Jakubowska, Z., Kulkova, I., Kowalczyk, P., Kramkowski, K., (2023). Biocontrol of fungal phytopathogens by *Bacillus pumilus*. *Front. Microbiol.* 14, 1194606.

Duchateau, S., Crouzet, J., Dorey, S., Aziz, A., (2024). The plant-associated *Pantoea* spp. as biocontrol agents: Mechanisms and diversity of bacteria-produced metabolites as a prospective tool for plant protection. *Biological Control*. 188, pp.105441.

EL Alami, N., EL Attari, S., (2020). Microbial biocontrol of post-harvest fungal rot in apples: Current State of the Science. *J. Botanical Res.* 2, 31-58.

Ezzati Yazid, S. N., Tajudin, N. I., Aribah Razman, N. A., Selamat, J., Ismail, S. I., Sanny, M., Putra Samsudin, N. I., (2023). Mycotoxigenic fungal growth inhibition and multi-mycotoxin reduction of potential biological control agents indigenous to grain maize. *Mycotoxin Res.* 39, 177-192.

Gotor-Vila, A., Teixidó, N., & Usall, J. (2017). Impact of volatile organic compounds from Bacillus subtilis strains on *Monilinia laxa* and *Botrytis cinerea* growth. *Journal of Applied Microbiology*, 123(2), 396–406.

Hipperdinger, M. L., Colman, D. I., Gortari, M. C., Pereyra, C. M., Astoreca, A. L., (2024). Characterization of *Epicoccum* isolates obtained from Argentinean sorghum grain samples. *Studies in Fungi.* 9, e003.

Khaskheli, M. A., Nizamani, M. M., Tarafder, E., Das, D., Nosheen, S., Muhae-Ud-Din, G., Khaskheli, R. A., Ren, M.-J., Wang, Y., Yang, S.-W., (2025). Sustainable management of major fungal phytopathogens in sorghum (*Sorghum bicolor L.*) for food security: A comprehensive review. *J. Fungi.* 11, 207.

Kurtzman, C. P., Robnett, C. J., (1998). Identification and phylogeny of ascomycetous yeasts from analysis of nuclear large subunit (26S) ribosomal DNA partial sequences. *Antonie van Leeuwenhoek*. 73, 331-371.

Lambrese, Y., Guiñez, M., Calvente, V., Sansone, G., Cerutti, S., Raba, J., Sanz, M. I. (2018). Production of siderophores by the bacterium *Kosakonia radicincitans* and its application to control of phytopathogenic fungi. *Biores. Technol. Reports*. 3, 82-87.

Leal Rodríguez, P., Rodríguez Galvão, J., da Silva Maia, B. K., Oliveira, C. A. M. M., Castro Zambrano, M. I., Barreira de Melo, N. F., (2024). Mechanisms of biocontrolling action of yeasts in post haervest fruit: a review. *Revista de Gestão Social e Ambiental* 18, 1-10.

Li, X., Yan, J., Li, D., Jiang, Y., Zhang, Y., Wang, H., Zhang, J., Ahmed, T., Li, B., (2021). Isolation and molecular characterization of Plant-Growth-Promoting bacteria and their effect on eggplant (*Solanum melongena*) growth. *Agriculture*. 11, 1258.

Mena Navarro, M. P., Espinosa Bernal, M. A., Martinez-Avila, A. E., Aponte Pineda, L. S., Montes Flores, L. A., Chan Ku, C. D., Hernández Gómez, Y. F., González Espinosa, J., Pacheco Aguilar, J. R., Ramos López, M. Á., Arvizu Gómez, J. L., Saldaña Gutierrez, C., Rodríguez Morales, J. A., Campos Guillén, J., (2024). Role of Volatile Organic Compounds Produced by *Kosakonia cowanii* Cp1 during Competitive Colonization Interaction against *Pectobacterium aroidearum* SM2. *Microorganisms*. 12, 930.

Mis, B., Karaca, K., Eltem, R., (2024). *In vitro* antagonistic activity of plant growth promoting rhizobacteria against aggressive biotypes of the green mold. *J. Basic Microbiol.* 64, e2400422.

Mohammed, A., Bekeko, Z., Yusufe, M., Sulyok, M., Krska, R., (2022). Fungal species and multi-mycotoxin associated with post-harvest sorghum (*Sorghum bicolor (L.) Moench*) grain in eastern Ethiopia. *Toxins.* 14, 473.

Navarta, L. G., Calvo, J., Posetto, P., Benuzzi, D., Sanz, M. I., (2020). Freeze-drying of a mixture of bacterium and yeast for application in postharvest control of pathogenic fungi. *SN App. Sci.* 2, 1223.

Oztekin, S., Dikmetas, D. N., Devecioglu, D., Acar, E. G., Karbancioglu-, F., (2023). Recent insights into the use of antagonistic yeasts for sustainable biomanagement of postharvest pathogenic and mycotoxigenic fungi in fruits with their prevention strategies against mycotoxins. *J. Agric. Food Chem.* 71, 9923-9950.

Ramírez-Pool, J. A., Calderón-Pérez, B., Ruiz-Medrano, R., (2024). Bacillus strains as effective biocontrol agents against phytopathogenic bacteria and promoters of plant growth. *Microbial Ecology.* 87, 76.

Schena, L., Ippolito, A., Zahavi, T., Cohen, L., Nigro, F., Droby, S., (1999). Genetic diversity and biocontrol activity of *Aureobasidium* pullulans isolates against postharvest rots. *Postharvest Biol. Technol.* 17, 189-199.

Taïbi, A., Meile, J.-C., Dieudonné, H., Korsten, L., Chillet, M. and Remize, F., (2020). New bacterial agents to limit *Colletotrichum gloeosporioides* development on mango. *Advances Microbiol.* 10, 691-712.

Toral, L., Rodríguez, M., Martínez-Checa, F., Montaño, A., Cortés-Delgado, A., Smolinska, A., Llamas, I., Sampedro, I., (2021). Identification of volatile organic compounds in extremophilic bacteria and their effective use in biocontrol of postharvest fungal phytopathogens. *Front. Microbiol.* 12, 773092.

Vasques, N. C., Nogueira, M. A., Hungria, M., (2024). Increasing application of multifunctional *Bacillus* for biocontrol of pests and diseases and plant growth promotion: lessons from Brazil. *Agronomy.* 14, 1654.

Vázquez-Gómez, E. I., Soria-Leal, L. Y., Chávez-Avilés M. N., (2019). Análisis del efecto antagonista de los compuestos orgánicios volátices (COV's) producidos por *Trichoderma* spp. y *Bacillus subtilis* de manera individual y en co-cultivo sobre el crecimiento de *Botrytis cinerea* causante de la pudrición de fresa (Capitulo 3). En: Aprovechamiento integral y calidad microbiológica de alimentos.

Vio, S. A., (2023). Microbiota bacteriana asociada al cultivo de tomate: caracterización y selección de bacterias promotoras del crecimiento vegetal. Tesis doctoral, Área Química, Universidad Nacional de La Plata.

Wang, Z., Han, Q., Lin, Y., & Chen, D. (2020). Microbial communities in sorghum seeds and their biocontrol potential. *Frontiers in Microbiology*, 11, 1417.

Zhao, X., Zhou., J, Tian R., Liu, Y., (2022). Microbial volatile organic compounds: antifungal mechanisms, applications, and challenges. *Microbiol.* 13, 922450.

Chapter 7

Fumonisins and Aflatoxins in the Maize Agri-Food Chain in Latin America: Biocontrol Approaches Using Fungal Enzymes

Marianela Bossa[1,*]
Noelía Edith Monesterolo[2]
Paloma Rhein[1]
Sofia Noemí Chulze[1]
María Silvina Alaniz Zanon[1]
and María Laura Chiotta[1,†]
[1]Instituto de Investigación en Micología y Micotoxicología (IMICO), Consejo Nacional de Investigaciones Científicas y Técnicas (CONICET) - Universidad Nacional de Río Cuarto (UNRC), Río Cuarto, Córdoba, Argentina
[2]Instituto de Biotecnología Ambiental y de la Salud (INBIAS), Consejo Nacional de Investigaciones Científicas y Técnicas (CONICET) - Universidad Nacional de Río Cuarto (UNRC), Río Cuarto, Córdoba, Argentina

Abstract

The prevalence of fumonisins and aflatoxins in the maize agri-food chain in Latin America, and particularly their impact on bioethanol production, is reviewed in this chapter. Maize (*Zea mays* L.) is a globally important crop, serving as a primary energy source for both humans and livestock. These mycotoxins, mainly produced by *Fusarium* and *Aspergillus* species, pose significant health risks to humans and animals. In addition

* Corresponding Author's Email: mbossa@exa.unrc.edu.ar
† Corresponding Author's Email: mchiotta@exa.unrc.edu.ar

In: Biocontrol Agents (BCAs)
Editor: Daiana García
ISBN: 979-8-90134-102-5

to its use as food and feed, maize is increasingly employed in bioethanol production, which is an essential renewable energy source. Several studies have shown that mycotoxins are resistant to the bioethanol production process and can even become more concentrated in the co-products intended for animal feed. Therefore, monitoring contamination levels throughout the bioethanol production process is critical to assess the associated mycotoxin risk in co-products. Various strategies have been explored for mycotoxin mitigation, with enzymatic biodegradation emerging as a particularly promising approach. Among the enzymes investigated, laccases stand out as one of the most effective candidates. This chapter highlights both the issue of mycotoxin contamination in Latin America and the challenge of implementing efficient mitigation strategies in industrial processes, to enhance the sustainability of maize-based systems and contribute to food and feed safety.

Keywords: maize, mycotoxins, bioethanol co-products, *Fusarium*, *Aspergillus*, Latin America

1. Introduction

Cereal crops such as wheat, maize, and rice are essential sources of sustenance for both humans and animals. Among them, maize is the most widely cultivated and consumed globally, with a production volume of 1,229,626 thousand metric tons recorded for the 2023/2024 season (Figure 1). During this period, Latin America accounted for approximately 16% of global maize production, with Brazil, Argentina, Mexico, and Colombia as the leading contributors. The main maize-exporting countries worldwide are the United States, Brazil, Argentina, and Ukraine (United States Department of Agriculture, USDA, 2025a,b).

Among cereal grains, maize (*Zea mays* L.) is one of the most efficient sources of metabolizable energy, primarily due to its high starch content, which includes approximately 72% of starch and low fiber levels. Although its protein concentration is relatively low, the large quantities in which maize is used make it a significant contributor of dietary protein in the livestock sector (Loy and Lundy, 2019).

The primary destination for maize produced in Argentina is grain exportation, which accounts for approximately 65% of the total production. Argentinian maize is mainly imported by countries in Southeast Asia, including Vietnam, Malaysia, and Indonesia; South America, particularly Chile and Peru; as well as Egypt, Saudi Arabia, Morocco, and Algeria (USDA,

2024a; USDA, 2025c; Ministerio de Economía de la República Argentina, MERA, 2025a). In contrast, Brazil's domestic market is the main destination for its maize production, with approximately 60 - 65% consumed internally, primarily for animal feed (poultry, swine, cattle) (60 - 70%). The remainder is allocated to human consumption and industrial applications, including maize flour, starch, oil, bioethanol, and various other products and derivatives (Empresa Brasileira de Pesquisa Agropecuária, Embrapa, 2024; USDA, 2025b). Despite high domestic demand, maize exports remain an important component of Brazil's agricultural economy. China is the leading importer, followed by Japan, South Korea, and Vietnam. Other major buyers of Brazilian maize include Spain, Portugal, Italy, the Netherlands, Iran, Saudi Arabia, Egypt, Algeria, and Morocco. Within Latin America, Mexico and Colombia are among Brazil's primary export markets (Embrapa, 2024; USDA, 2024b). Although Mexico exports only a small percentage of its maize grain, it is the leading producer of maize for direct human consumption in Latin America (Guzzon et al., 2021).

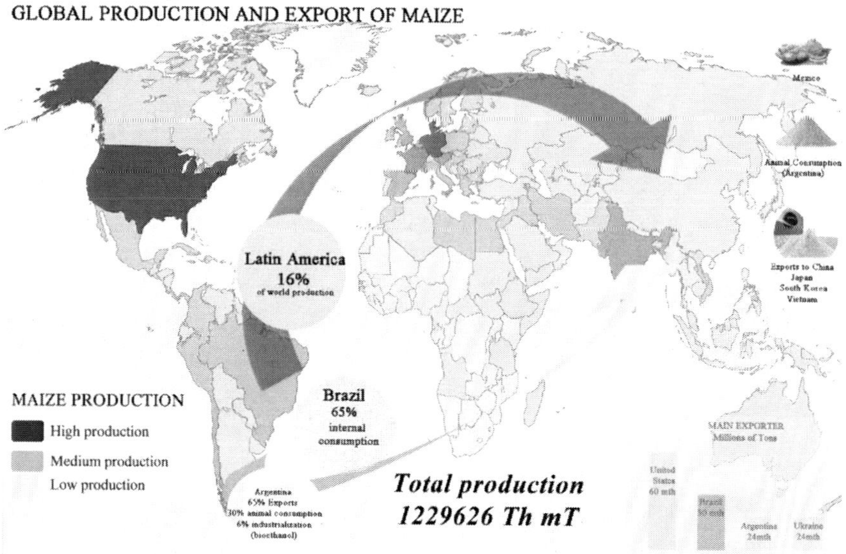

Figure 1. Production and exportation of maize grains from Latin American countries, and their ranking worldwide. Th mT: Thousand metric tons.

In Argentina, approximately 6% of maize production is directed towards industrial uses, including dry milling, wet milling, and seed production.

Additionally, about 4% is utilized for bioethanol production (Bolsa de Comercio de Rosario, BCR, 2025; MERA, 2025a). In 2024, Brazil was the second-largest producer of maize-based ethanol globally (28%), while Argentina ranked eighth, contributing 1% to global production (Renewable Fuels Association, RFA, 2025). In Argentina, approximately 60% of the bioethanol produced in 2024 was maize-based, with the remaining portion derived from sugarcane (MERA, 2025b). In contrast, in Brazil, sugarcane is the primary raw material for ethanol production, followed by maize. In 2024, production shares were estimated at 80% from sugarcane and 19% from maize (USDA, 2024c). In Mexico, the high production costs of bioethanol significantly limit its development. In countries such as Costa Rica and Colombia, the main feedstocks for bioethanol production include sugarcane, sorghum, and coffee pulp (Vega et al., 2024).

A common phytosanitary issue in maize production is the colonization of grains by various fungal species, which often leads to contamination with mycotoxins, posing significant risks to both human and animal health. Among the primary mycotoxins produced by *Fusarium* and *Aspergillus* species, fumonisins (FBs), aflatoxins (AFs), deoxynivalenol, zearalenone, and ochratoxin A are frequently detected in maize and maize-derived products (Munkvold et al., 2019). Contamination levels depend on several factors, including meteorological conditions during the growing season, storage and processing practices, and the presence and prevalence of toxigenic fungi, among many others (Kolawole et al., 2021; Roucou et al., 2022; Zheng et al., 2024). In this context, the development and implementation of effective preventive and control strategies require the monitoring of region-specific mycotoxin contamination levels throughout the different stages of the maize production chain.

This contamination not only compromises the safety of maize-based foods but also affects animal-derived products. Animals exposed to mycotoxin-contaminated feed can accumulate these toxins in their tissues, milk, and eggs, thereby posing a secondary risk to human health. Furthermore, these metabolites are not eliminated during conventional food processing or industrial manufacturing (Stracquadanio et al., 2021).

Fumonisins are primarily synthesized by *Fusarium verticillioides* and *F. proliferatum*, both members of the *Fusarium fujikuroi* species complex, as well as by certain species within the *Aspergillus* section *Nigri* (Wigmann et al., 2020; Logrieco et al., 2021; Dinolfo et al., 2022). In addition to their high mycotoxin-producing potential, these species are also associated with maize ear rot (Wigmann et al., 2020; Logrieco et al., 2021; Dinolfo et al., 2022).

Fumonisins have been linked to equine leukoencephalomalacia, hepatic and renal toxicity in rodents, porcine pulmonary edema, and neural tube defects in humans (Knutsen et al., 2018). Among them, fumonisin B_1 (FB_1) is considered the most relevant in terms of both toxicity and occurrence (Voss et al., 2007). Due to its toxicity, FB_1 has been classified by the International Agency for Research on Cancer (IARC) as a Group 2B carcinogen (IARC, 2002).

Aflatoxins (AFs) are also highly relevant mycotoxins in terms of toxicity and food safety. *Aspergillus flavus* and *A. parasiticus*, both members of the *Aspergillus* section *Flavi*, are the primary producers of AFs (Wigmann et al., 2020; Camardo Leggieri et al., 2021; Dinolfo et al., 2022). Among all the AFs, four types are the most widely recognized and frequently detected in food: AFB_1, B_2, G_1, and G_2, with AFB_1 standing out due to its high toxicity and prevalence. Exposure to AFs can result in various adverse effects, including hepatotoxicity, immunotoxicity, and teratogenicity (Bennett and Klich, 2003; Dhakal et al., 2023). Aflatoxin B_1 is considered the most toxic naturally occurring compound and has been classified by IARC as a Group 1 human carcinogen (2002).

Given the risks associated with mycotoxin contamination, many national agencies have established maximum permissible limits for individual mycotoxins in a wide range of products and by-products. In Latin America, Mercosur does not set maximum limits for FBs in maize, although its member states do. For example, in Argentina, the Código Alimentario Argentino (CAA) establishes a maximum level of 2,000 μg/kg for total FBs (FB_1 + FB_2) in maize flour and maize grits (CAA, 2025). In Brazil, Agência Nacional de Vigilância Sanitária (ANVISA) has set a maximum limit of 5,000 μg/kg for total FBs (FB_1 + FB_2) in maize intended for further processing (ANVISA, 2011). Regarding AFs, the maximum allowable for AFB_1 is 5 μg/kg, and 20 μg/kg for total AFs in maize and its by-products (Mercosur, 2002).

This chapter provides a review of the current situation regarding FB and AF contamination in harvested maize grains in Latin America, as well as in maize by-products, particularly the co-products generated during the bioethanol production process from maize. In addition, it offers an updated review of various mycotoxin mitigation strategies applied to maize grains and by-products, with a particular emphasis on the use of fungal enzymes, which represent one of the most recent and promising tools developed in recent years.

2. Contamination of Maize Grains with Fumonisins and Aflatoxins

Since mycotoxin contamination in cereals is influenced by multiple factors, such as climate, storage conditions, and the presence of various fungal species, the levels detected can vary significantly across countries and regions. The following section provides a detailed overview of FB and AF contamination levels in maize grains from Latin American countries.

2.1. Fumonisins

Fumonisins are frequently detected in maize (Bryla et al., 2022; Gomes et al., 2024). In Latin America, contamination with FBs is widespread, with reported levels varying across countries and regions.

In Argentina, natural mycotoxin occurrence in maize was monitored over 10 years. A total of 3,246 samples of freshly harvested (n = 1,655) and stored (n = 1,591) maize were collected from various regions of the country (Table 1). Throughout the survey, 90 to 100% of the samples were contaminated with FB_1, with mean levels ranging from 268.8 to 3,956.4 μg/kg in freshly harvested maize, and from 877.5 to 6,497 μg/kg in stored maize (Garrido et al., 2012). Industrial wet-milling plants located in different regions (Northwest, Center, Southeast, and Southwest) provided 43 maize samples for routine FBs analysis. All samples were contaminated with FBs, with concentrations ranging from 244 to 14,864 μg/kg (Coronel et al., 2016). Another study, 90 maize samples collected from commercial fields in Buenos Aires province over multiple harvest seasons, showed no detectable levels of either FB_1 or FB_2 (Castañares et al., 2019). Moschini et al. (2020) analyzed FBs concentrations in 58 maize grain samples, detecting levels as high as 17,944 μg/kg. Another study analyzed 72 raw maize samples from five provinces, identifying both free and hydrolyzed forms of FBs, with contamination levels ranging from below the detection limit to 8,388 μg/kg, highlighting the widespread presence of these mycotoxins in Argentina (Pok et al., 2020). Maize silo bags prepared in Buenos Aires, Santa Fe, and Córdoba were also monitored during storage. Fumonisins levels increased by 27% from pre- to post-storage, from 3,079.1 to 4,238.5 μg/kg, respectively (Pacin et al., 2009). Additionally, studies on infant cereals containing maize by-products showed an alarming 84% contamination rate, with levels ranging from 0.05 to 992 μg/kg. Maize puffs, a popular snack, had a mean FBs concentration of

257.5 μg/kg, with some samples exceeding 1,649 μg/kg (Federico et al., 2010; Cendoya et al., 2024).

In Brazil, both the frequency and concentration of FBs in maize and maize-based food have been reported to be high. Total FBs in maize samples from Minas Gerais State were evaluated, and all samples were found to be contaminated with levels ranging from 230 to 6,450 μg/kg (Queiroz et al., 2012). Another survey in Rio Grande do Sul State analyzed 29 maize samples, reporting contamination with FB_1 in 58.6% of samples, FB_2 in 37.9%, and both in 37.9% (Stumpf et al., 2013). In Santa Catarina State, 232 maize samples intended for commercialization were collected. Fumonisins were detected in 46.6% of the samples, with concentrations ranging from 66 to 7,832 μg/kg for FB_1 and from 110 to 1,201 μg/kg for FB_2 (Scussel et al., 2014). Oliveira et al. (2017) analyzed 148 maize samples from fields in the states of Rio Grande do Sul, Santa Catarina, and Paraná. Fumonisin B_1 and B_2 were detected in 100% of the samples, while FB_3, FB_4, and FB_6 were found in 98.6, 97.9, and 8.1 percent of the samples, respectively. Total FBs (FB_1 + FB_2) levels ranged from 63.8 to 66,274 μg/kg in Paraná, from 62.4 to 10,080 μg/kg in Santa Catarina, and from 120.5 to 24,581 μg/kg in Rio Grande do Sul States. In more recent studies, de Matos et al. (2024) reported that all maize-based samples analyzed were contaminated with FBs, with levels ranging from 7.2 to 2,158.2 μg/kg. Additionally, hidden FBs were detected in all samples, with concentrations between 19.2 and 750.8 μg/kg, often exceeding free FBs.

In a study carried out in Peru, FBs were analyzed in 82 maize grain samples. Total FBs were detected in 63.4% of the samples, with levels ranging from 125 to 27,490 μg/kg and a mean of 1,651.1 μg/kg (Coloma et al., 2019).

In Venezuela, few studies have investigated the natural occurrence of FBs in maize. A study carried out in the states of Yaracuy and Guárico reported FBs contamination in all white maize grain samples collected, particularly those obtained from major production areas. Fumonisin levels ranged from 400 to 5,100 μg/kg and from 300 to 72,000 μg/kg in samples collected from Yaracuy and Guárico States, respectively (Chavarri et al., 2017).

Table 1. Natural incidence of fumonisin in maize grains collected from different Latin American countries

Country	Region	Mycotoxins	Mycotoxin-positive simples/total samples	Range (μg/kg)	Mean (μg/kg)	Method detection	References
Argentina	Several regions	FBs	319/326	10 – 31,108	3,658.8	HPLC	Pacin et al. (2009)
	Several regions	FBs	3169/3246	333.5 – 8,881	3,133.2	HPLC	Garrido et al. (2012)
	Industrial plants	FBs	43/43	244 – 14,864	5,154.9	HPLC	Coronel et al. (2016)
		FB_1	43/43	132 – 10,352	3,506.72		
	Buenos Aires Province		0/90	–	–	HPLC- MS/MS	Castañares et al. (2019)
	Pampa region		58	ND – 17,944	NI	HPLC and HPLC-MS/MS	Moschini et al. (2020)
	Buenos Aires Province	FBs	32/36	12 – 3,833	778	HPLC-FLD	Pok et al. (2020)
	Córdoba Province		2/2	483 – 1,026	755		
	Entre Ríos Province		7/12	6 – 2,515	436		
	Río Negro Province		1/1	–	5,215		
	Santa Fe Province		21/21	719 – 8,387	2,810		
Brazil	Minas Gerais State	FBs	40/40	230 – 6,450	2,338.5	FumoniTest immunochemical method (VICAM, USA)	Queiroz et al. (2012)
	Río Grande do Sul State	FB_1	17/29	90 – 2,030	660	HPLC	Stumpf et al. (2013)
		FB_2	11/29	130 – 840	420		
	Santa Catarina State	FB_1	108/232	66 – 7,832	1,114	HPLC	Scussel et al. (2014)
		FB_2	25/232	110 – 1,201	438		

Country	Region	Mycotoxins	Mycotoxin-positive simples/total samples	Range (µg/kg)	Mean (µg/kg)	Method detection	References
Brazil	Rio Grande do Sul State	FBs	51/51	120.5 – 24,581	2,726	LC-MS/MS	Oliveira et al. (2017)
	Santa Catarina State		40/40	62.4 – 10,080	2,204		
	Paraná State		57/57	63.8 – 66,274	3,153		
	Several regions		212/212	7.2 – 2,158.2		UHPLC-MS/MS	de Matos et al. (2024)
Peru	Commercial establishments	FBs	52/82	125 – 27,490	1,651.1	LC-MS/MS	Coloma et al. (2019)
Venezuela	Guárico State	FBs	11/11	300 – 72,000	8,400	FumoniTest immunochemical method (VICAM, USA)	Chavarri et al. (2017)
	Yaracuy State		11/11	400 – 5,100	1,800		
Guatemala	22 departments	FB_1	633/640	ND – 17,100	1,800	LC-MS/MS	Torres et al. (2015)
	Huehuetenango Department	FBs	51/98	400 – 31,000	NI	Lateral quantifiable enzyme-linked immunosorbent assay (Agra-Strip, Romer Labs, Union, MO)	Mendoza et al. (2018)
Mexico	Jalisco State	FBs	6/36	100 – 700	430	ELISA (AgraQuant® test kit, Romer Labs.)	Reyes-Velazquez et al. (2008)
	Puebla State	FBs	12/21	90 – 580	341	Ridascreen ELISA kit (R-Biopharm, Germany)	Peña Betancourt and Posadas Peña (2018)
	Mexico City		9/10	220 – 551	375		
	Mexico City	FB_1	63/64	79 – 1,589	488	LC-MS/MS	Gilbert-Sandoval et al. (2020)
		FB_2	64/64	24 – 524	225		

NI: Not informed.
ND: Not detected.

In Guatemala, 640 maize samples intended for human and/or animal consumption were collected from various markets across 22 departments. Nearly 99% of the samples tested positive for FB_1, with concentrations ranging from below the detection limit to 17,100 μg/kg. Samples from the departments of Chiquimula and Jutiapa were the most heavily contaminated, with concentrations ranging from 200 to 13,200 μg/kg (mean 3,200 μg/kg) and from 200 to 17,100 μg/kg (mean 3,100 μg/kg), respectively. Samples from the departments of Chimaltenango and Escuintla also showed high contamination levels, ranging from 20 to 5,000 μg/kg (mean 700 μg/kg) and from 20 to 13,500 μg/kg (mean 2,700 μg/kg), respectively (Torres et al., 2015). In Chiantla and Todos Santos in the department of Huehuetenango, FBs were analyzed in 98 samples from 25 farms. Total FBs (FB_1, FB_2, and FB_3) were detected in 52% of the samples, with concentrations ranging from 400 to 31,000 μg/kg (Mendoza et al., 2018).

Mexico is a country with high maize consumption. In a survey conducted by Peña Betancourt and Posadas Peña (2019), a total of 31 raw maize samples were collected from Mexico City and eight localities in Puebla State for FB_1 analysis. Contamination was detected in 90% and 57% of the samples from Mexico City and Puebla, respectively, with FB_1 concentrations ranging from 200 to 551 μg/kg and 90 to 580 μg/kg. In another study, FB_1 and FB_2 were detected in 64 nixtamalized maize samples from Mexico City, with contamination levels up to 1,589 μg/kg of FB_1 and 524 μg/kg for FB_2 (Gilbert-Sandoval et al., 2020). Additionally, stored maize samples contained several mycotoxins, including FBs, with concentrations ranging from 100 to 700 μg/kg (Reyes-Velazquez et al., 2008).

In summary, FBs contamination varies considerably across Latin American countries. In Brazil, contamination is widespread, with high levels detected in several states, particularly in the southern regions, where climatic factors such as humidity and temperature contribute to increased contamination (dos Santos Lopes et al., 2023). In Argentina, contamination rates are frequently high, with 90 - 100% of maize samples testing positive, often at elevated concentrations, especially in stored maize, where high moisture content and prolonged storage periods exacerbate the issue (Pacin et al., 2009). Similarly, Guatemala and Mexico reported high contamination levels. Venezuela has also documented the presence of FBs, particularly in key maize-producing states. Differences in FBs occurrence are influenced by regional agricultural practices, climatic conditions, harvest seasons, and storage methods, with post-harvest handling playing a critical role in exacerbating contamination. Moreover, these factors may contribute to the cooccurrence of multiple mycotoxins. For instance, FBs and AFs are often found together in essential foods (Martínez et al., 2004).

2.2. Aflatoxins

According to several reports, AF levels are increasing in various crops, including maize and its derivatives (Bhardwaj et al., 2022). In Latin America, countries such as Argentina, Brazil, Colombia, Venezuela, and Uruguay have conducted studies on mycotoxins in maize, with a particular focus on AFs (Ponce-García et al., 2021).

In Entre Ríos, Argentina, the mean recorded AF levels were 3.19 μg/kg in 2003 and 2.0 μg/kg in 2004 (Broggi et al., 2007) (Table 2). Additionally, in the previously mentioned survey conducted by Garrido et al. (2012), contamination with AFB_1 was reported in 5.86% of freshly harvested maize samples and 10.5% of stored maize samples. Aflatoxin B_1 concentrations ranged from 0.38 to 2.54 μg/kg in fresh maize and from 0.22 to 4.5 μg/kg in stored samples. In the semi-arid region of Chaco, Argentina, *A. flavus* is the main species responsible for AFs contamination in maize. A study by Barontini et al. (2024) identified 28 aflatoxigenic isolates and found a positive correlation between the number of sclerotia and AFB_1 production, highlighting significant mycotoxin risks for maize crops. González Pereyra et al. (2008) investigated *Aspergillus* species and AFB_1 contamination in pre- and post-fermented maize silage. Aflatoxin B_1 was not detected in any pre-fermented silo samples. However, some samples from the upper and lower sections of post-fermented silos showed low-level contamination (<10 μg/kg). In contrast, six samples from the middle section showed AFB_1 levels ranging from 1.43 to 155.78 μg/kg. *Aspergillus* spp. was the most frequently isolated fungal genus (78%), with *A. flavus* being the most prevalent species (53%).

A study carried out on 18 commercial farms in the south-central and eastern regions of Uruguay found that maize silage, a commonly used livestock feed in the country, showed significant contamination with AFs. Specifically, two out of 34 samples exceeded the maximum recommended level of 20 μg/kg established by Uruguay's Ministry of Livestock, Agriculture, and Fisheries (Capelli et al., 2019).

Table 2. Natural incidence of aflatoxin in maize grains collected from different Latin America countries

Country	Region	Mycotoxins	Mycotoxin-positive simples/total samples	Range (µg/kg)	Mean (µg/kg)	Method detection	References
Argentina	Entre Ríos Province	AFs	14/31	ND – 22.4	2.6	TLC	Broggi et al. (2007)
	San Luis and Córdoba Provinces	AFB_1	6/60	ND – 155.78	NI	HPLC	González-Pereyra et al. (2008)
	Several regions	AFB_1	264/3246	ND – 427	2.6	HPLC	Garrido et al. (2012)
Uruguay	Canelones, Lavalleja, and San José Departments		22/22*	2.2 – 44.1	9.5	ELISA	Capelli et al. (2019)
Brazil	Paraná State	AFB_1	12/74	ND – 3	< 0.8	LC-MS/MS	de Souza et al. (2013)
	Paraná State	AFB_1	9/16	ND – 97.4	28.88	HPLC	Viaro et al. (2017)
	Rio Grande do Sul State	AFs	23/51	0.7 – 35.1	7.8	LC-MS/MS	Oliveira et al. (2017)
	Santa Catarina State		4/40	0.5 – 49.9	16.37		
	Paraná State		11/57	0.4 – 16.7	3.1		
	São Paulo and Santa Catarina States	AFB_1	1/26	NI	3.6	HPLC-MS/MS	Franco et al. (2019)
	Paraná State	AFB_1	22/39	0.11 – 3.91	1.51	ic-ELISA	Silva et al. (2022)
Peru	Commercial establishments	AFs	53/82	1 – 17	2.1	LC-MS/MS	Coloma et al. (2019)
Colombia		AFB_1	8/8	20.9 – 30.6	26.3	HPLC	Díaz et al. (2004)
	Atlantic Coast	AFB_1	11/11	15.2 – 282.6	NI	HPLC	Acuña et al. (2005)
	Five departments of the Colombian Caribbean	AFB_1	3/24	2.4 – 12.5	7.9	HPLC	Morris Navarro (2011)

Country	Region	Mycotoxins	Mycotoxin-positive simples/total samples	Range (µg/kg)	Mean (µg/kg)	Method detection	References
Venezuela	Sabana de Parra, Agua Blanca, Santa Cruz de Turén and San Javier	AFs	31/36	ND – 121.08	17.9	Specific immunoassay and/or ELISA	Chavarri et al. (2009)
Costa Rica	Feed industries	AFs	246/1321	ND – 420	NI	ELISA and HPLC	Granados-Chinchilla et al. (2017)
Guatemala	All the 22 departments	AFB_1	206/572	0 – 2,656	63	-	Torres et al. (2015)
Haiti	Port au-Prince and Nord Department	AFs	6/30	ND – 78	< 9.6	Immuno-affinity column chromatography and fluorometric detection	Schwartzbord and Brown (2015)
	Sud Department	AFs	7/9	ND – 1,500	NI	ELISA	Aristil et al. (2017)
	Sud Department	AFB_1	11/20	NI	172.4	HPLC	Aristil et al. (2020)
Mexico	Puebla State	AFs	16/21	2 – 7.24	3.62	HPLC	Peña Betancourt and Posadas Peña (2019)
	Mexico City		6/10	2 – 7.8	6.89		

ND: Not detected.

NI: Not informed.

*These samples include not only maize but also sorghum kernels. The authors did not discriminate between both kinds of samples when presenting the results.

In a study carried out in Paraná State, Brazil, AFs were evaluated in 74 maize grain samples. Aflatoxin B_1 was detected in 16% of the samples, with concentrations ranging from below the detection limit to 3 µg/kg. Aflatoxin G_2 was also detected in three maize samples ranging from not detected to 1.8 µg/kg and AFG_1 in one sample at a level of 0.6 µg/kg (de Souza et al., 2013). Viaro et al. (2017) analyzed 16 maize grain samples from the same state and found total AFs (B_1 + B_2 + G_1 + G_2) in 56% of the samples, with concentrations ranging from 10.23 to 191.63 µg/kg. This percentage was higher than that reported by Oliveira et al. (2017), who found AFs in 19.3% of maize samples from Paraná State. In that survey, AFB_1 was present in all positive samples, and even higher levels were observed in samples from Santa Catarina and Rio Grande do Sul States. Notably, the total AFs concentrations in some cases exceeded the maximum permitted limit established by ANVISA (2011). More recent studies conducted in northern Paraná reported that 56.4% of maize samples were positive for AFs, with a mean contamination level of 1.51 µg/kg (Silva et al., 2022). In contrast, a study in Piauí found no AFs in maize flour samples intended for human consumption, despite the presence of potentially mycotoxin-producing fungi (Cardoso Filho et al., 2020). However, a study done by Franco et al. (2019) revealed median AFB_1 levels of 3.6 µg/kg in maize-based food and up to 100 µg/kg in animal feed from small-scale farms.

In Peru, 82 samples of purple maize were analyzed. Total AFs were detected in 64.6% of the samples, with concentrations ranging from 1 to 17 µg/kg and a mean of 1.9 µg/kg (Coloma et al., 2019).

Aflatoxins, particularly AFB_1, are prevalent in Colombian maize, with contamination levels often exceeding the international standard of 10 - 20 µg/kg (Díaz et al., 2004). Morris Navarro (2011) investigated AFs levels in cereals such as rice and maize intended for human consumption across five departments of the Colombian Caribbean. Samples were randomly collected from factories and distributors in different cities. Aflatoxin B_1 was detected in 12.5% of the maize samples at levels between 2.4 and 12.5 µg/kg (mean of 7.9 µg/kg). Aflatoxin B_2 was found in 8.3% of the samples, at concentrations below 2 µg/kg, while AFG_1 and G_2 were not detected. The mean total AFs concentration was 9.2 µg/kg. Another survey conducted along the Atlantic coast reported a 100% incidence of AFB_1 in maize, with levels exceeding 100 µg/kg. This contamination affected 50,000 hectares and resulted in a 35% reduction in yield (Acuña et al., 2005).

In Venezuela, AFs were detected in white maize harvested under irrigation in the states of Yaracuy and Portuguesa between 2005 and 2007, with several samples exceeding the regulatory limit of 20 μg/kg (Chavarri et al., 2009). Additionally, the authors demonstrated that the contamination was specifically associated with *A. flavus*.

In Costa Rica, as part of a national surveillance program, feedstuff samples including maize were analyzed. The maximum concentration AFs detected was 72.14 μg/kg. In the same study, maize samples intended for human consumption were also analyzed. The prevalence of AFs in white and yellow maize was 38.6 and 7.5%, respectively, with maximum concentrations of 420 and 410 μg/kg, respectively (Granados-Chinchilla et al., 2017).

In Guatemala, AFB_1 was analyzed in 572 maize grain samples, and 36% tested positive, with values up to 2,656 μg/kg. Samples that were visibly infected with fungi showed significantly higher AFB_1 levels compared to visually clean samples (Torres et al., 2015).

In Haiti, Schwartzbord and Brown (2015) analyzed maize grain samples obtained from Port-au-Prince and Nord Department. No AFs were detected in the samples from Port-au-Prince, whereas those from Nord Department showed a 30% incidence of AFs, with concentrations ranging from below the detection limit to 78 μg/kg. Aristil et al. (2017) evaluated the occurrence of toxigenic fungi and AFs contamination in maize grains collected from the Sud Department. Aflatoxins were detected in 75% of the samples, with values ranging from not detected to 1,500 μg/kg. In another study, these authors evaluated AFs contamination in the same area and detected that 55% of the maize samples were contaminated with AFB_1, with a mean concentration of 172.4 ± 282.5 μg/kg. These AFs concentrations exceed both the European (4 μg/kg) and American (20 μg/kg) regulatory limits for human food in all samples, indicating a potential chronic exposure of the Haitian population to AFs (Aristil et al., 2020).

A study carried out in Mexico revealed that 76% of maize samples from Puebla State were contaminated with AFs, with concentrations ranging from 2 to 7.24 μg/kg. Additionally, 60% of the maize samples collected in Mexico City contained AFs, with levels ranging from 2 to 7.8 μg/kg (Peña Betancourt and Posadas Peña, 2019).

To summarize, Brazil, Argentina, and Colombia reported widespread AFs contamination, with high levels detected in multiple regions. In Venezuela, contamination appears to be associated with irrigation practices, while Uruguay shows comparatively lower levels, particularly in maize intended for human consumption. Guatemala, Mexico, Peru, and Venezuela also reported notable contamination, especially in specific regions. In Costa Rica, higher AFs levels have been detected in animal feed, whereas in Haiti, significant contamination has been documented, with marked regional variability. In most of the studies conducted across Latin America, mycotoxin concentrations in maize frequently exceeded the regulatory limits established by national and international entities, raising serious concerns regarding the potential health risks to consumers. Moreover, monitoring and research on mycotoxin contamination remain limited in many of these countries, underscoring the need for continuous surveillance and the implementation of more robust analytical methods to improve the understanding of AFs prevalence. These efforts are essential to ensure food safety, particularly in light of the chronic exposure risk faced by populations that regularly consume contaminated maize.

3. Bioethanol Production Process from Maize and Mycotoxin Contamination in Co-Products

Bioethanol is produced through the fermentation of various feedstocks rich in fermentable sugars or carbohydrates, and the process is generally categorized into four main generations. First-generation bioethanol is primarily derived from food crops such as maize, wheat, barley, sugarcane, and various vegetable oils, which provide high concentrations of fermentable sugars that enhance production efficiency (Halder et al., 2019). In contrast, second-generation bioethanol is obtained from non-edible raw materials, including lignocellulosic biomass from forestry residues, woody and herbaceous plants, non-food crops, municipal solid waste, and animal fats (Devi et al., 2022). Third-generation bioethanol is produced from algal biomass, offering potential advantages in terms of yield and environmental sustainability. Fourth-generation bioethanol is synthesized using captured carbon dioxide as a raw material through advanced technologies such as electrochemical conversion and fourth-generation hydroprocessing techniques (Azad et al., 2015, 2016; Azad, 2017; Halder et al., 2019). Although the industrial production of bioethanol from high-sucrose crops like sugarcane or starch-rich crops such as maize and wheat is well-established, the control of AFs and FBs in maize-based bioethanol production remains a significant challenge.

In Latin America, the risk of mycotoxin contamination is severe (World Health Organization, WHO, 2023). According to the Global Mycotoxin Survey, maize is highly contaminated, with mean concentrations of 2,050 and 16 ppb for FBs and AFs, respectively (DSM-Firmenich, 2024). The issue worsens when contaminated maize is used for bioethanol production, as the fermentation process can concentrate FBs and AFs up to threefold in co-products such as Wet Distillers Grains (WDGs), Dry Distillers Grains (DDGs), and other animal feed alternatives (Mohammadi et al., 2021; Buenavista et al., 2021) (Figure 2). Several studies have demonstrated a high prevalence of FBs in distillers' grains, often exceeding the limits established by the Food and Drug Administration (FDA). In contrast, only about 3% of samples analyzed for AFs in this matrix surpass the regulatory thresholds. These findings underscore the potential health risk associated with the consumption of such products by livestock (Magnoli et al., 2019; Li et al., 2024). Moreover, the presence of mycotoxins has been shown to induce stress in the yeast strains employed during fermentation, negatively affecting ethanol production yields (Liu, 2011; Pinotti et al., 2016).

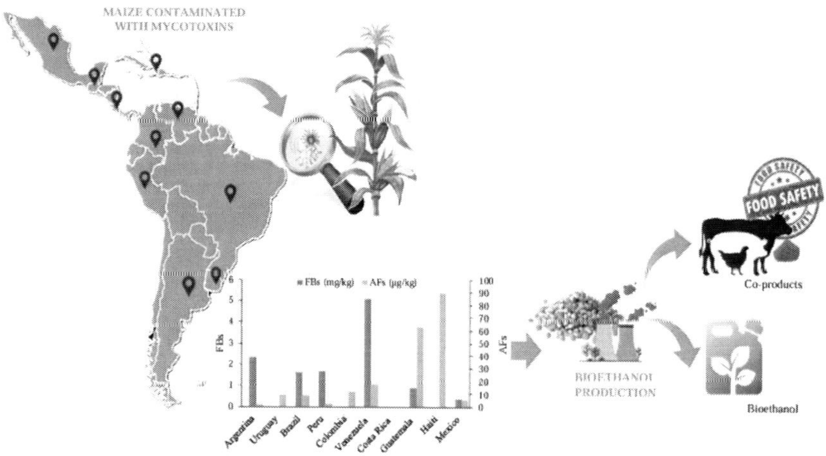

Figure 2. Maize contamination with mycotoxins in Latin America. Several countries have reported contamination with fumonisins and aflatoxins both in maize grains and by-products of this crop, including products intended for feed and food, and co-products obtained during the bioethanol production process.

3.1. Occurrence of Mycotoxins in the Bioethanol Production Process from Maize

During the different stages of bioethanol production, such as liquefaction, saccharification, fermentation, and distillation, mycotoxins may undergo transformation, adsorption, or degradation, all of which influence their final concentrations in the resulting co-products. Understanding the behavior of mycotoxins throughout the production process is essential for evaluating the safety of these co-products when used as animal feed and for developing effective mitigation strategies.

Few studies have investigated mycotoxin contamination across the different stages of the bioethanol production process. Dzuman et al. (2016) analyzed the presence of mycotoxins in samples collected throughout the process, including input maize, 11 intermediate products, the final bioethanol, and DDGs. Solid samples included maize, maize flour, DDGs, and WDGs, while liquid samples were taken during liquefaction, saccharification, fermentation (first and second stages), distillation (whole stillage and bioethanol), and evaporation (condensed distillers' solubles, CDS or "syrup"). Fumonisin B_1 levels increased significantly during liquefaction, saccharification, and fermentation, reaching up to 269 μg/kg. This observation contrasts with the findings of Bothast et al. (1992), who reported a decrease in FBs during fermentation. In the stillage samples, FB_1 levels further increased, with most of the toxin transferring to the CDS fraction obtained by centrifugation and evaporation. In WDGs, FB_1 concentrations were lower compared to stillage, whereas in DDGs, the final dried product, FB_1 levels were approximately double those found in the input maize.

Bossa et al. investigated the natural occurrence of AFs and FB_1 in various intermediates of the bioethanol production process in Argentina. Samples included maize grains at the process input, maize steep liquor after the saccharification-liquefaction stage, maize steep liquor at the end of fermentation, whole stillage, thin stillage following centrifugation, and the final co-product, WDG. Aflatoxins were not detected in any of the samples; however, FB_1 was present in concentrations ranging from 150 to 8,640 μg/kg (unpublished data). The FB_1 levels increased notably during the liquefaction and fermentation stages. Similar observations were reported by Dzuman et al. (2016), who attributed the increase to the enzymatic breakdown of matrix components by amylase, which releases previously bound mycotoxins and increases their measurable concentrations in the by-products. Additionally, extracellular enzymes secreted by yeasts during fermentation may further contribute to the rise of FB_1 levels. In this study, FB_1 was predominantly transferred into the WDGs fraction, with concentrations approximately 1.5 times higher than those measured in the input maize across all three batches. These findings are consistent with those of Buenavista et al. (2021), reinforcing the notion that mycotoxin concentrations typically increase throughout the bioethanol production process.

3.2. Natural Incidence of Mycotoxins in DDGs

The contamination of bioethanol co-products with mycotoxins, particularly in DDGs, has become an increasing concern due to its implications for animal feed safety. Given that mycotoxins tend to concentrate during the bioethanol production process, it is essential to evaluate their prevalence and levels in DDGs to assess the potential risks associated with their inclusion in livestock diets. Numerous studies have investigated the natural occurrence of mycotoxins in DDGs, offering valuable insights into their distribution patterns, regulatory compliance, and potential health risks for animals.

Although numerous studies have been conducted on mycotoxin contamination in DDGs, there is a lack of research in Latin America. A recent study conducted in Brazil analyzed 186 DDGs samples from a bioethanol plant and found a high prevalence of FB_1 and FB_2, detected in 98.8% and 97.6% of the samples, respectively. The mean concentrations of FB_1 and FB_2 were 3,207 μg/kg and 1,243 μg/kg. Aflatoxin B_1 (AFB_1) was detected in 32.3% of the samples, with a mean concentration of 1.47 μg/kg (Mallmann et al., 2021). In Costa Rica, a 29% prevalence of AFs contamination in DDGs samples was reported, with a maximum concentration of 86.8% (Granados-Chinchilla et al., 2017).

Studies in other countries have evaluated mycotoxin contamination in DDGs. For instance, in the United States, Zhang et al. (2009) detected FBs in 98.35% of samples (n= 182), with concentrations ranging from 100 to 8,600 μg/kg, while AFB_1 was found in 6.04% of samples, at levels between 1.04 and 3.7 μg/kg. A follow-up study confirmed the presence of FBs in 100% of samples collected from ethanol plants, port containers, and their destination in Taiwan, with concentrations ranging from 900 to 2,300 μg/kg. Similarly, Zhang and Caupert (2012) reported FBs in 89.5% of 67 samples from United States ethanol plants, with concentrations ranging from not detected to 8,900 μg/kg, and AFs in 19.4% of samples, with levels between 1 and 5,7 μg/kg. Rodrigues and Chin (2012) analyzed 409 DDGs samples from the United States, Asia, and Europe, detecting FBs in 91% of samples with a mean concentration of 1,036 μg/kg, and AFs in 19% of samples with a mean of 2 μg/kg. Additional studies by Khatibi et al. (2014) and Munkvold et al. (2019) further confirmed these trends, finding FBs and AFs among the most prevalent mycotoxins in DDGs globally, with contamination reported in samples from the United States ethanol plants and other regions. These findings highlight the widespread issue of mycotoxin contamination in DDGs across diverse regions. Brazilian studies have similarly reported a high prevalence of FBs, consistent with observations in the United States and Europe, although data on the presence of other mycotoxins, such as AFs, remain limited. The scarcity of research across Latin America continues to hinder a comprehensive understanding of mycotoxins prevalence in the region.

4. Mycotoxin Mitigation Strategies in Maize and by-Products

The mitigation of mycotoxin accumulation in maize grains can be achieved through the implementation of various integrated strategies at different stages of the production chain. These include preventive measures applied in the field (pre-harvest), during storage, and at the processing stage (post-harvest), all aiming at minimizing mycotoxins presence and ensuring food safety. Proper implementation of pre-harvest control methods is critical for effective mycotoxin prevention in cereals, nuts, spices, and legumes, thereby reducing their entry into the post-harvest stage (Magan and Aldred, 2007). However, pre-harvest strategies alone are insufficient to guarantee the complete absence of mycotoxins in food products (Karlovsky et al., 2016). The main post-harvest strategies for mycotoxin reduction focus on controlling grain moisture content, water activity, and temperature during storage and transportation (Nada et al., 2022). Nevertheless, these measures alone are not adequate, highlighting the need for additional reduction or decontamination strategies during industrial processing. These may include mycotoxin elimination, biotransformation, decontamination, and detoxification (Alaniz Zanon et al., 2025).

Several studies have explored the use of physical, chemical, enzymatic, and microbial methods during food and feed processing (Urugo et al., 2023; Agriopoulou et al., 2025). The most well-known and widely applied physical methods include grain sorting and sieving, milling, maceration and extrusion, thermal treatment, irradiation, and the use of adsorbents (Baglieri et al., 2013; Çelik et al., 2013; Peivasteh-Roudsari et al., 2022). Recently, novel physical technologies for mycotoxin decontamination in food have emerged, including cold plasma, pulsed light, pulsed electric field, high-pressure processing, and electron beam irradiation (Urugo et al., 2023). Mycotoxins can be transformed into less toxic or non-toxic compounds through chemical approaches such as peroxidation, ammoniation, the use of bisulfites, acid or alkali treatments, and the application of oxidizing agents like ozone (Nguyen et al., 2020; Nunes et al., 2021; Agriopoulou et al., 2025). To enhance the overall efficacy, physical and chemical methods are often combined (Urugo et al., 2023). Nevertheless, the application of these mycotoxin detoxification strategies in food and feed is limited due to their relatively low efficacy, high costs, and potential adverse effects on the quality of raw material. Additionally, these methods may generate toxic residues in food and the environment, while also modifying organoleptic properties and nutritional composition of the final products (Gavahian and Cullen, 2020; Alaniz Zanon et al., 2025; Agriopoulou et al., 2025).

In the last decade, biological detoxification methods have gained increasing interest as alternatives to physical and chemical strategies for the post-harvest elimination of mycotoxins, owing to their low cost, reduced toxicity, and minimal impact on food quality. Several living microorganisms, including bacteria, fungi, and yeasts, have demonstrated the ability to metabolize, degrade, transform, deactivate, or destroy mycotoxins. Similarly, decontamination can be achieved through the use of enzymes. These biological strategies are capable of converting mycotoxins into stable, less toxic, or even harmless compounds (Loi et al., 2017; Peivasteh-Roudsari et al., 2022; Urugo et al., 2023).

Different microorganism species have demonstrated the ability to degrade AFs and FBs. For instance, in the case of AFs degradation, bacterial genera such as *Pseudomonas* (Samuel et al., 2014; Sangare et al., 2014; Adebo et al., 2016; Xu et al., 2023), *Rhodococcus* (Risa et al., 2018; Deng et al., 2022), *Streptomyces* (Campos-Alvelar et al., 2021), and *Bacillus* (Shi et al., 2024) have been reported as effective AFs degraders, achieving reductions of up to 100%. Regarding FBs degradation, a group of bacteria composed of *Pseudomonas, Delftia, Sphingobacterium, Achromobacter*, and other species has degraded up to 90% of FB_1 (Zhao et al., 2019). Additionally, *Lactobacillus brevis* has been reported not only to degrade FB_1 but also to adsorb it, achieving a 90% reduction in maize silages (Martínez Tuppia et al., 2017).

Mycotoxin-degrading fungi have also been reported. Several strains of *Saccharomyces cerevisiae* have demonstrated the ability to degrade AFB_1 and FB_1 by up to 53% and 78%, respectively (Armando et al., 2013; Chlebicz and Śliżewska, 2020; Guan et al., 2021; Qu et al., 2022). Additionally, filamentous fungi have shown high efficiency in degrading these mycotoxins. For instance, *Exophiala spinifera* is capable of utilizing FB_1 as its sole carbon source (Blackwell et al., 1999), and *Aspergillus welwitschiae* and *A. niger* have biotransformed FB_1 (Qi et al., 2016; Garnham et al., 2020). In the case of AFs, certain non-toxigenic *Aspergillus* species, such as *A. niger*, which is widely used in the food industry, are capable of decontaminating AFB_1 by up to 95% (Zhang et al., 2014; Sun et al., 2016). While the degradation mechanisms in some of these studies remain unclear, in others the reduction is attributed to adsorption. This is considered one of the primary mechanisms of mycotoxin detoxification, alongside enzymatic degradation (Qu et al., 2022; Sandlin et al., 2022). In this context, the use of fungal enzymes for mycotoxin decontamination holds strong potential for industrial-scale application, as will be described in the following section.

4.1. Enzymatic Biotransformation of Mycotoxins

The use of microorganisms for mycotoxin decontamination is infrequent due to constraints such as low degradation efficiency, potential alterations in the sensory properties of the final product, and possible toxicity (Wang and Xie, 2020). Consequently, recent research has increasingly focused on the application of enzymes for mycotoxin decontamination, with promising potential in both the food and feed industries. Several enzymes have been characterized for their ability to catalyze the degradation of one or more mycotoxins. These enzymes participate in various metabolic pathways involved in the breakdown of FBs or AFBs, leading to the formation of less toxic or non-toxic degradation products. Among the main enzyme classes with activity against AFs are oxidoreductases, lactonases, and other specialized enzymes such as the aflatoxin-degrading enzyme from *Pantoea* and *Pseudomonas* (Xie et al., 2019; Song et al., 2019; Cheng et al., 2023). The most extensively studied oxidoreductases include oxidases, reductases, peroxidases, and laccases (Das and Mishra, 2000; Alberts et al., 2009; Yehia, 2014; Li et al., 2019; Loi et al., 2016; Bossa et al., 2024a). Among these, AFs-oxidase (AFO), produced by *Armillariella tabescens*, was the first enzyme identified with the capacity to completely degrade AFB_1 (Liu et al., 1998; Cao et al., 2011; Wu et al., 2015). For FB_1 degradation, numerous enzymes, including carboxylesterases, amine oxidases, aminotransferases, manganese peroxidases and laccases, have demonstrated high degradation capacities, with reductions up to 100% in some cases (Heinl et al., 2010, 2011; Hartinger and Moll, 2011; Loi et al., 2018; Wang et al., 2019; Bossa et al., 2024b; Li et al., 2024). Currently, only one enzyme has been approved by the European Food Safety Authority (EFSA) for FB_1 degradation in pig feed, which is known as FUMzyme® (Biomin Holding GmbH, Getzersdorf, Austria) (EFSA, 2014). This commercial enzyme is a FumD esterase (EC 3.1.1.87) derived from *Sphingopyxis* sp. MTA144 and produced by a genetically modified strain of *Komagataella pastoris*. It is considered safe for humans, animals, and the environment (Heinl et al., 2010; Alberts et al., 2019; Li et al., 2024). Other enzymes, such as laccases, have also shown efficacy in the detoxification of both FBs and AFs.

4.1.1. Laccase Enzymes

Laccases (EC 1.10.3.2) are enzymes classified within the multicopper oxidase family (Aza and Camarero, 2023). These enzymes are capable of oxidizing a broad spectrum of compounds, both aromatic and non-aromatic. Their low substrate specificity provides them with remarkable catalytic versatility, allowing the degradation or transformation of numerous compounds of interest across environmental, industrial, and biotechnological applications (Agrawal et al., 2018). Laccases are synthesized by a variety of organisms, including plants, insects, bacteria, and fungi. However, those produced by white-rot fungi within the phylum Basidiomycota are particularly noteworthy due to their high efficiency in lignin degradation, and as such, they are the most extensively studied (Rodríguez-Couto et al., 2019).

To further enhance the catalytic efficiency of laccases, the use of redox mediators has been reported. These mediators facilitate electron transfer between the enzyme and the substrate, thereby expanding the enzyme's range of activity to include compounds that are otherwise difficult to oxidize (Ashe et al., 2016; Mani et al., 2018). The strong oxidative capacity of laccases, combined with their versatility, has led to their increasing application in several industries, including pulp and paper, textiles, bioremediation, organic synthesis, and the food industry (Bossa et al., 2024b).

Besides their industrial applications, laccases have been extensively studied for their ability to degrade various mycotoxins. Alberts et al. (2009) evaluated the degradation of AFB_1 using a purified laccase from *Trametes versicolor* (1 U/mL), achieving an 87% reduction in the concentration of the mycotoxin. Since then, numerous studies have reported laccase synthesis by *T. versicolor*, *Pleurotus ostreatus*, *P. pulmonaris*, and *P. eryngii*, and these enzymes demonstrated AFs decontamination efficiencies up to 76%. Additionally, these studies confirmed the enhanced efficacy of redox mediators such as acetosyringone. However, most of these experiments were conducted in buffer systems rather than in food or feed matrices (Zeinvand-Lorestani et al., 2015; Loi et al., 2016; Loi et al., 2018). Notably, Bossa et al. (2024a) evaluated a laccase synthesized by *T. gallica* B7-IMICO-UNRC and achieved a 26% reduction of AFB_1 in maize steep liquor samples derived from the liquefaction stage of the bioethanol production process.

Regarding FB's degradation by laccases, studies to date remain limited. Loi et al. (2018) evaluated the degradative capacity of the Ery4 laccase from *P. eryngii* against FB_1, achieving a maximum reduction of 74% using TEMPO (2,2,6,6-tetramethylpiperidinyloxyl) as a redox mediator. Bossa et al. (2024b) assessed the efficacy of several Basidiomycota strains for FB_1 degradation, under conditions both with and without redox mediators (vanillic acid and TEMPO), and using maize steep liquor as a matrix. The *T. gallica* B4-IMICO-RC strain exhibited the highest FB_1 degradation efficiency (62.2%). In this study, the presence of redox mediators did not enhance the degradation percentages.

The strategy to decontaminate FBs and AFs focused on maize steep liquor obtained from the bioethanol production process could serve as a promising tool. Nevertheless, further validation is needed. Ultimately, the goal of this research is to apply a laccase during the fermentation stage to decontaminate mycotoxins and produce safe DDGs intended for animal feed.

The applicability of enzymes for mycotoxin detoxification in food and feed depends largely on the toxicity of the resulting degradation products, as these compounds may pose risks equal to or even greater than those of the original mycotoxins, potentially threatening human and animal health (Guan et al., 2021; Li et al., 2024; Alaniz Zanon et al., 2025). Loi et al. (2023) demonstrated that the Ery 4 laccase transforms AFB_1 to AFQ_1, epi-AFQ_1, AFB_1-8,9-epoxide, AFB_1 dialdehyde, AFB_2a, AFM_1, and P1. These findings suggest that the resulting biodegradation products are either non-toxic or exhibit reduced carcinogenic potential in humans. Bossa et al. also demonstrated that degradation products of AFs obtained using enzymatic extracts containing laccases were not cytotoxic to human colorectal adenocarcinoma cells (Caco-2 cell line) (unpublished data). Consequently, the use of enzymatic extracts represents a promising and effective strategy for the decontamination of food products contaminated with AFs (Kumar et al., 2021).

Conclusion and Perspectives

Throughout this chapter, a comprehensive survey of the occurrence of fumonisins and aflatoxins in maize and maize-derived products in Latin America was conducted, with a particular emphasis on co-products from the bioethanol industry, intended for animal feed. The presence of these mycotoxins can significantly affect the health of animals consuming contaminated feed, resulting in economic losses for producers. Furthermore, if such contaminated products enter the human food chain and reach the final consumer, they may pose a risk to the population's health. Although several strategies have been developed over time to mitigate the effects of these mycotoxins, no method currently ensures their complete elimination from food or feed. The selection of an appropriate mycotoxin control strategy should consider economic and technical feasibility, safety, and the ability to preserve nutritional value without generating toxic compounds or hazardous residues. Currently, one of the most promising technologies is the use of enzymes to decontaminate these toxins, converting them into less toxic or even innocuous metabolites. Continued scientific and technological advancements, alongside coordinated efforts, will facilitate the implementation of enzymatic detoxification not only in the bioethanol production industry from maize but also across other sectors related to the production of raw materials and foods derived from maize and other cereals and grains. Understanding the mechanisms by which these enzymes interact with their substrates, as well as optimizing their integration into specific production processes, will be key to expanding their applicability and enhancing food safety. Additionally, while enzymatic detoxification of mycotoxins presents a promising approach, the absence of an established legal framework governing their decontamination in food products remains a significant barrier. Establishing such regulations is crucial to ensure food and feed safety.

Acknowledgments

The authors thank Dr. César Horacio Casale from INBIAS (Instituto de Biotecnología Ambiental y Salud; UNRC-CONICET) of Río Cuarto, Córdoba, Argentina, for his important contribution to this work.

Funding

This study was supported by grants from CONICET through different subsidies: PIBAA (2022-1930-APN-DIR#CONICET), PIP CONICET (11220200100560CO, Res. 2021-1639-APN-DIR# CONICET), and PUE (22920200100004CO, Res. 2020-1672-APN-DIR#CONICET); from Agencia de Investigación, Desarrollo e Innovación del Fondo para la Investigación Científica y Tecnológica through PICT (2022-03-00456).

References

Acuña, C. A., Díaz, G. J., & Espitia, M. E. (2005). Aflatoxinas en maíz: reporte de caso en la Costa Atlántica Colombiana. *Revista de la Facultad de Medicina Veterinaria y de Zootecnia*, *52*(2), 156–162.

Adebo, O. A., Njobeh, P. B., Sidu, S., Tlou, M. G., & Mavumengwana, V. (2016). Aflatoxin B_1 degradation by liquid cultures and lysates of three bacterial strains. *International Journal of Food Microbiology*, *233*, 11–19. https://doi.org/10.1016/j.ijfoodmicro.2016.06.007.

Agência Nacional de Vigilância Sanitária (ANVISA) (2011). Brazilian Sanitary Surveillance Agency: Resolução n° 7, de 18 de fevereiro de 2011.

Agrawal, K., Chaturvedi, V., & Verma, P. (2018). Fungal laccase discovered but yet undiscovered. *Bioresources and Bioprocessing*, *5*, 4. https://doi.org/10.1186/s40643-018-0190-z.

Agriopoulou, S., Smaoui, S., & Varzakas, T. (2025). Postharvest management of mycotoxins in supply chains of agricultural products. *CABI Review*, *20*(1), 0011. https://doi.org/10.1079/cabireviews.2025.0011.

Alaniz Zanon, M. S., Bossa, M., Chiotta, M. L., & Chulze, S. N. (2025). Biological strategies for aflatoxin mitigation in food and feed chains in a changing world. In J. C. Assaf (Ed.), *Battling Aflatoxins- Advances in Food and Feed Research [Working Title]*. IntechOpen. https://doi.org/10.5772/intechopen.1009018.

Alberts, J. F., Gelderblom, W. C. A., Botha, A., & van Zyl, W. H. (2009). Degradation of aflatoxin B_1 by fungal laccase enzymes. *International Journal of Food Microbiology*, *135*(1), 47–52. https://doi.org/10.1016/j.ijfoodmicro.2009.07.022.

Alberts, J., Schatzmayr, G., Moll, W. D., Davids, I., Rheeder, J., Burger, H. M., Shephard, G., & Gelderblom, W. (2019). Detoxification of the fumonisin mycotoxins in maize: An enzymatic approach. *Toxins*, *11*(9), 523. https://doi.org/10.3390/toxins11090523.

Aristil, J., Venturini, G., & Spada, A. (2017). Occurrence of toxigenic fungi and aflatoxin potential of *Aspergillus* spp. strains associated with subsistence farmed crops in Haiti. *Journal of Food Protection*, *80*(4), 626–631. https://doi.org/10.4315/0362-028X.JFP-16-278.

Aristil, J., Venturini, G., Maddalena, G., Toffolatti, S. L., & Spada, A. (2020). Fungal contamination and aflatoxin content of maize, moringa and peanut foods from rural subsistence farms in South Haiti. *Journal of Stored Products Research*, *85*, Article 101550. https://doi.org/10.1016/j.jspr.2019.101550.

Armando, M. R., Galvagno, M. A., Dogi, C. A., Cerrutti, P., Dalcero, A. M., & Cavaglieri, L. R. (2013). Statistical optimization of culture conditions for biomass production of probiotic gut-borne Saccharomyces cerevisiae strain able to reduce fumonisin B_1. *Journal of Applied Microbiology*, *114*(5), 1338–1346. https://doi.org/10.1111/jam.12144.

Ashe, B., Nguyen, L. N., Hai, F. I., Lee, D. J., van de Merwe, J. P., Leusch, F. D. L., Price, W. E., & Nghiem, L. D. (2016). Impacts of redox-mediator type on trace organic contaminants degradation by laccase: Degradation efficiency, laccase stability and effluent toxicity. *International Biodeterioration and Biodegradation*, *113*, 169–176. https://doi.org/10.1016/j.ibiod.2016.04.027.

Aza, P., & Camarero, S. (2023). Fungal laccases: Fundamentals, engineering and classification update. *Biomolecules*, *13*(12), 1716. https://doi.org/10.3390/biom13121716.

Azad, A. K. (2017). Biodiesel from mandarin seed oil: A surprising source of alternative fuel. *Energies*, *10*(11), 1689. https://doi.org/10.3390/en10111689.

Azad, A. K., Rasul, M. G., Khan, M. M. K., Sharma, S. C., & Hazrat, M. A. (2015). Prospect of biofuels as an alternative transport fuel in Australia. *Renewable and Sustainable Energy Reviews*, *43*, 331–351. https://doi.org/10.1016/j.rser.2014.11.047.

Azad, A. K., Rasul, M. G., Khan, M. M. K., Sharma, S. C., Mofijur, M., & Bhuiya, M. M. K. (2016). Prospects, feedstocks and challenges of biodiesel production from beauty leaf oil and castor oil: A nonedible oil sources in Australia. *Renewable and Sustainable Energy Reviews*, *61*, 302–318. https://doi.org/10.1016/j.rser.2016.04.013.

Baglieri, A., Reyneri, A., Gennari, M., & Nègre, M. (2013). Organically modified clays as binders of fumonisins in feedstocks. *Journal of Environmental Science and Health, Part B*, *48*(9), 776–783. https://doi.org/10.1080/03601234.2013.780941.

Barontini, J., Alaniz Zanon, M. S., Torrico, A. K., Druetta, M., Luna, I. M., Posse, A. R., Chulze, S. N., & Giménez Pecci, M. P. (2024). A polyphasic study of non-aflatoxigenic *Aspergillus flavus* Link, isolates from maize in the Chaco semi-arid region of Argentina. *Revista de La Facultad de Ciencias Agrarias, Universidad Nacional de Cuyo*, *56*(1), 58–73. https://www.scielo.org.ar/scielo.php?script=sci_arttext&pid=S1853-86652024000100058&lng=es&nrm=iso&tlng=en.

Bennett, J. W., & Klich, M. (2003). Mycotoxins. *Clinical Microbiology Reviews*, *16*(3), 497–516. http://dx.doi.org/10.1128/CMR.16.3.497-516.2003.

Bhardwaj, K., Adunphatcharaphon, S., Banerjee, K., Elliott, C., Petchkongkaew, A., & Kolawole, O. (2022). A review of the fundamental factors and processes leading to the accumulation of aflatoxins In Cereal Crops. *Prepints*. https://doi.org/10.20944/preprints202201.0400.v1.

Blackwell, B. A., Gilliam, J. T., Savard, M. E., David Miller, J., & Duvick, J. P. (1999). Oxidative deamination of hydrolyzed fumonisin B_1 (AP1) by cultures of *Exophiala spinifera*. *Natural Toxins*, *7*(1), 31–38. https://doi.org/10.1002/(SICI)1522-7189(199902)7:1%3C31::AID-NT36%3E3.0.CO;2-W.

Bolsa de Comercio de Rosario (BCR). (2025, February 28). *Maíz: Balance de Oferta y Demanda en Argentina*. https://www.bcr.com.ar/es/mercados/investigacion-y-desarrollo/informativo-semanal/estadisticas-informativo-semanal/maiz-97.

Bossa, M., Alaniz Zanon, M. S., Monesterolo, N. E., Monge, M. del P., Coria, Y. M., Chulze, S. N., & Chiotta, M. L. (2024a). Aflatoxin decontamination in maize steep liquor obtained from bioethanol production using laccases from species within the Basidiomycota Phylum. *Toxins*, *16*(1), 27. https://doi.org/10.3390/toxins16010027.

Bossa, M., Monesterolo, N. E., Monge, M. del P., Rhein, P., Chulze, S. N., Alaniz-Zanon, M. S., & Chiotta, M. L. (2024b). Fungal laccases and fumonisin decontamination in co-products of bioethanol from maize. *Toxins*, *16*(8), 350. https://doi.org/10.3390/toxins16080350.

Bothast, R. J., Bennett, G. A., Vancauwenberge, J. E., & Richard, J. (1992). Fate of fumonisin B_1 in naturally contaminated corn during ethanol fermentation. *Applied and

Environmental Microbiology, *58*(1), 233–236. https://doi.org/10.1128/aem.58.1.233-236.1992.

Broggi, L. E., Pacin, A. M., Gasparovie, A., Sacchi, C., Rothermel, A., Gallay, A., & Resnik, S. (2007). Natural occurrence of aflatoxins, deoxynivalenol, fumonisins and zearalenone in maize from Entre Ríos Province, Argentina. *Mycotoxin Research*, *23*(2), 59–64. https://doi.org/10.1007/BF02946026.

Bryla, M., Pierzgalski, A., Zapaśnik, A., Uwineza, P. A., Ksieniewicz-Woźniak, E., Modrzewska, M., & Waśkiewicz, A. (2022). Recent research on *Fusarium* mycotoxins in maize – A review. *Foods*, *11*(21), 3465. https://doi.org/10.3390/foods11213465.

Buenavista, R. M. E., Siliveru, K., & Zheng, Y. (2021). Utilization of distiller's dried grains with solubles: A review. *Journal of Agriculture and Food Research*, *5*, Article 100195. https://doi.org/10.1016/j.jafr.2021.100195.

Camardo Leggieri, M., Toscano, P., & Battilani, P. (2021). Predicted aflatoxin B_1 increase in Europe due to climate change: Actions and reactions at global level. *Toxins*, *13*(4), 292. https://doi.org/10.3390/toxins13040292.

Campos-Avelar, I., de la Noue, A. C., Durand, N., Cazals, G., Martinez, V., Strub, C., Fontana, A., & Schorr-Galindo, S. (2021). *Aspergillus flavus* growth inhibition and aflatoxin B_1 decontamination by *Streptomyces* isolates and their metabolites. *Toxins*, *13*(5), 340. https://doi.org/10.3390/toxins13050340.

Cao, H., Liu, D., Mo, X., Xie, C., & Yao, D. (2011). A fungal enzyme with the ability of aflatoxin B_1 conversion: Purification and ESI-MS/MS identification. *Microbiological Research*, *166*(6), 475–483. https://doi.org/10.1016/j.micres.2010.09.002.

Capelli, A., Suárez, G., & García y Santos, C. (2019). Aflatoxinas en alimentos y leche de vacas de 18 establecimientos comerciales de las regiones centro-sur y este de Uruguay. *Veterinaria (Montevideo)*, *55*(212), 52–56. https://doi.org/10.29155/vet.55.212.2.

Cardoso Filho, F. C., Calvet, R. M., Pereyra, C. M., Pereira, M. M. G., Rosa, C. A. R., Torres, A. M., & Muratori, M. C. S. (2020). Occurrence of *Aspergillus* spp., *Penicillium* spp. and aflatoxins in corn flour samples used in human consumption, Piauí, Brazil. *Arquivos Do Instituto Biológico*, *78*(3), 443–447. https://doi.org/10.1590/1808-1657v78p4432011.

Castañares, E., Martínez, M., Cristos, D., Rojas, D., Lara, B., Stenglein, S., & Dinolfo, M. I. (2019). *Fusarium* species and mycotoxin contamination in maize in Buenos Aires province, Argentina. *European Journal of Plant Pathology*, *155*(4), 1265–1275. https://doi.org/10.1007/s10658-019-01853-5.

Çelik, K., Uzatıcı, A., Coşkun, B., & Demir, E. (2013). Current developments in removal of mycotoxins by biological methods and chemical adsorbents. *Journal of Hygienic Engineering and Design*, 17–20. https://keypublishing.org/jhed/wp-content/uploads/2020/07/Full-paper-Kemal-Celik.pdf.

Cendoya, E., Nichea, M. J., Romero Donato, C. J., Zachetti, V. G. L., Monge, M. P., & Ramirez, M. L. (2024). Fumonisins in infant cereals marketed as complementary food in Argentina. *Food Additives & Contaminants: Part B*, *18*(1), 48–54. https://doi.org/10.1080/19393210.2024.2411278.

Chavarri, M., Barroyeta, J., Ochoa, Y., Rumbos, N., Alezones, J., Chavarri, M., Barroyeta, J., Ochoa, Y., Rumbos, N., & Alezones, J. (2017). Detección de *Fusarium verticillioides* y fumonisinas en granos de maíz blanco provenientes de los estados

Yaracuy y Guárico, Venezuela. *Nova Scientia*, *9*(19), 171–184. https://doi.org/10.21640/ns.v9i19.1035.

Chavarri, M., Luzón, O., Mazzani, C., González, C., Alezones, J., & Garrido, M. J. (2009). Mohos toxigénicos y micotoxinas en maíz de grano blanco cosechado bajo riego en los estados Yaracuy y Portuguesa, Venezuela. *Fitopatología Venezolana*, *22*(1), 2–7.

Cheng, S., Wu, T., Zhang, H., Sun, Z., Mwabulili, F., Xie, Y., Sun, S., Ma, W., Li, Q., Yang, Y., Wu, X., & Jia, H. (2023). Mining lactonase gene from aflatoxin B_1-Degrading strain *Bacillus megaterium* and degrading properties of the recombinant enzyme. *Journal of Agricultural and Food Chemistry*, *71*(51), 20762–20771. https://doi.org/10.1021/acs.jafc.3c05725.

Chlebicz, A., & Śliżewska, K. (2020). *In vitro* detoxification of aflatoxin B_1, deoxynivalenol, fumonisins, T-2 toxin and zearalenone by probiotic bacteria from genus *Lactobacillus* and *Saccharomyces cerevisiae* yeast. *Probiotics and Antimicrobial Proteins*, *12*(1), 289–301. https://doi.org/10.1007/s12602-018-9512-x.

Código Alimentario Argentino (CAA). 2025. *Capítulo III: De los productos alimenticios.* https://www.argentina.gob.ar/sites/default/files/capitulo_iii_prod_alimenticiosactualiz_2025-04.pdf.

Coloma, Z. N., Oliveira, M. S., Dilkin, P., Mallmann, A. O., Almeida, C. A. A., & Mallmann, C. A. (2019). Mycotoxin occurrence in Peruvian purple maize. *World Mycotoxin Journal*, *12*(3), 307–316. https://doi.org/10.3920/WMJ2019.2436.

Coronel, M. B., Vicente, S., Resnik, S. L., Alzamora, S. M., & Pacin, A. (2016). Fumonisins in maize and gluten meal analysed in Argentinean wet milling industrial plants by ELISA compared with HPLC-FLD method. *Food Control*, *67*, 285–291. https://doi.org/10.1016/j.foodcont.2016.02.052.

Das, C., & Mishra, H. N. (2000). *In vitro* degradation of aflatoxin B_1 by horse radish peroxidase. *Food Chemistry*, *68*(3), 309–313. https://doi.org/10.1016/S0308-8146(99)00196-X.

de Matos, N. A. V., Sartori, A. V., de Sá Soilo, E. R. P., de Moraes, M. H. P., & Jacob, S. C. (2024). A survey on free and hidden fumonisins in Brazilian corn and corn-based products. *Food Control*, *156*, 110135. https://doi.org/10.1016/j.foodcont.2023.110135.

de Souza, M. L. M., Sulyok, M., Freitas-Silva, O., Costa, S. S., Brabet, C., Machinski Junior, M., Sekiyama, B. L., Vargas, E. A., Krska, R., & Schuhmacher, R. (2013). Cooccurrence of mycotoxins in maize and poultry feeds from Brazil by Liquid Chromatography/Tandem Mass Spectrometry. *The Scientific World Journal*, *2013*, Article 427369. http://dx.doi.org/10.1155/2013/427369.

Deng, D., Tang, J., Liu, Z., Tian, Z., Song, M., Cui, Y., Rong, T., Lu, H., Yu, M., Li, J., Pang, R., & Ma, X. (2022). Functional characterization and whole-genome analysis of an aflatoxin-Degrading *Rhodococcus pyridinivorans* strain. *Biology*, *11*(5), 774. https://doi.org/10.3390/biology11050774.

Devi, A., Bajar, S., Kour, H., Kothari, R., Pant, D., & Singh, A. (2022). Lignocellulosic biomass valorization for bioethanol production: A circular bioeconomy approach. *Bioenergy Research*, *15*(4), 1820–1841. https://doi.org/10.1007/s12155-022-10401-9.

Dhakal, A., Hashmi, M. F., & Sbar, E. (2023). Aflatoxin toxicity. *In StatPearls Publishing.*

Díaz, G. J., Perilla, N. S., & Espitia, E. (2004). First Colombian interlaboratory study for the determination of aflatoxin B_1 in yellow corn. *Mycotoxin Research*, *20*(1), 11–18. https://doi.org/10.1007/BF02946703.

Dinolfo, M. I., Martínez, M., Castañares, E., & Arata, A. F. (2022). *Fusarium* in maize during harvest and storage: a review of species involved, mycotoxins, and management strategies to reduce contamination. *European Journal of Plant Pathology*, *164*(2), 151–166. https://doi.org/10.1007/s10658-022-02548-0.

dos Santos Lopes, L., Ramos, G. L. P. A., Alvarenga, V. O., Campagnollo, F. B., Oliveira, S. B. S., Freire, L., & Sant'Ana, A. S. (2023). Quantitative risk assessment of the presence of fumonisin in corn produced in different regions of Brazil: Influence of climatic conditions. *Environmental Research*, *237*, 116907. https://doi.org/10.1016/j.envres.2023.116907.

dsm-firmenich. (2024). *World Mycotoxin Survey. The Global Threat January – December 2023*. https://www.dsm-firmenich.com/content/dam/dsm/anh/en/documents/REP_MTXsurvey_Q4_2023_EN_0124_AUE_doublePage.pdf.

Dzuman, Z., Stranska-Zachariasova, M., Vaclavikova, M., Tomaniova, M., Veprikova, Z., Slavikova, P., & Hajslova, J. (2016). Fate of free and conjugated mycotoxins within the production of distiller's dried grains with solubles (DDGS). *Journal of Agricultural and Food Chemistry*, *64*(24), 5085–5092. https://doi.org/10.1021/acs.jafc.6b00304.

EFSA Panel on Additives and Products or Substances used in Animal Feed (FEEDAP). (2014). Scientific Opinion on the safety and efficacy of fumonisin esterase (FUMzyme®) as a technological feed additive for pigs. *EFSA Journal*, *12*(5), 3667.

Empresa Brasileira de Pesquisa Agropecuária (Embrapa). (2024, May 24). *Artigo: Dia Nacional do Milho - A Importância do Milho para o Agronegócio Brasileiro*. https://www.embrapa.br/busca-de-noticias/-/noticia/89583335/artigo-dia-nacional-do-milho---a-importancia-do-milho-para-o-agronegocio-brasileiro.

Federico, F. J., Funes, G. J., Salas, M. P., & Resnik, S. L. (2010). Natural occurrence of fumonisins in corn puff in Argentina. *Mycotoxin Research*, *26*(4), 273–278. https://doi.org/10.1007/s12550-010-0066-3.

Franco, L. T., Petta, T., Rottinghaus, G. E., Bordin, K., Gomes, G. A., & Oliveira, C. A. F. (2019). Co-occurrence of mycotoxins in maize food and maize-based feed from small-scale farms in Brazil: a pilot study. *Mycotoxin Research*, *35*(1), 65–73. https://doi.org/10.1007/s12550-018-0331-4.

Garnham, C. P., Butler, S. G., Telmer, P. G., Black, F. E., Renaud, J. B., & Sumarah, M. W. (2020). Identification and characterization of an *Aspergillus niger* amine oxidase that detoxifies intact fumonisins. *Journal of Agricultural and Food Chemistry*, *68*(47), 13779–13790. https://doi.org/10.1021/acs.jafc.0c04504.

Garrido, C. E., Hernández Pezzani, C., & Pacin, A. (2012). Mycotoxins occurrence in Argentina's maize (*Zea mays* L.), from 1999 to 2010. *Food Control*, *25*(2), 660–665. https://doi.org/10.1016/j.foodcont.2011.11.043.

Gavahian, M., & Cullen, P. J. (2020). Cold plasma as an emerging technique for mycotoxin-free food: Efficacy, mechanisms, and trends. *Food Reviews International*, *36*(2), 193–214. https://doi.org/10.1080/87559129.2019.1630638.

Gilbert-Sandoval, I., Wesseling, S., & Rietjens, I. M. C. M. (2020). Occurrence and probabilistic risk assessment of fumonisin B_1, fumonisin B_2 and deoxynivalenol in nixtamalized maize in Mexico City. *Toxins*, *12*(10), 644. https://doi.org/10.3390/toxins12100644.

Gomes, A. L., Petrus, R. R., de Sousa, R. L. M., & Fernandes, A. M. (2024). Aflatoxins and fumonisins in conventional and organic corn: a comprehensive review. *Food Additives & Contaminants: Part A*, *41*(5), 575–586. https://doi.org/10.1080/19440049.2024.2330092.

González Pereyra, M. L., Alonso, V. A., Sager, R., Morlaco, M. B., Magnoli, C. E., Astoreca, A. L., Rosa, C. A. R., Chiacchiera, S. M., Dalcero, A. M., & Cavaglieri, L. R. (2008). Fungi and selected mycotoxins from pre- and postfermented corn silage. *Journal of Applied Microbiology*, *104*(4), 1034–1041. https://doi.org/10.1111/j.1365-2672.2007.03634.x.

Granados-Chinchilla, F., Molina, A., Chavarría, G., Alfaro-Cascante, M., Bogantes-Ledezma, D., & Murillo-Williams, A. (2017). Aflatoxins occurrence through the food chain in Costa Rica: Applying the One Health approach to mycotoxin surveillance. *Food Control*, *82*, 217–226. https://doi.org/10.1016/j.foodcont.2017.06.023.

Guan, Y., Chen, J., Nepovimova, E., Long, M., Wu, W., & Kuca, K. (2021). Aflatoxin detoxification using microorganisms and enzymes. *Toxins*, *13*(1), 46. https://doi.org/10.3390/toxins13010046.

Guzzon, F., Rios, L. W. A., Cepeda, G. M. C., Polo, M. C., Cabrera, A. C., Figueroa, J. M., Hoyos, A. E. M., Calvo, T. W. J., Molnar, T. L., León, L. A. N., León, T. P. N., Kerguelén, S. L. M., Rojas, J. G. O., Vázquez, G., Preciado-Ortiz, R. E., Zambrano, J. L., Rojas, N. P., & Pixley, K. V. (2021). Conservation and use of Latin American maize diversity: Pillar of nutrition security and cultural heritage of humanity. *Agronomy*, *11*(1), 172. https://doi.org/10.3390/AGRONOMY11010172.

Halder, P., Azad, K., Shah, S., & Sarker, E. (2019). Prospects and technological advancement of cellulosic bioethanol ecofuel production. In K. Azad (Ed.), *Advances in Eco-Fuels for a Sustainable Environment* (pp. 211–236). Woodhead Publishing. https://doi.org/10.1016/B978-0-08-102728-8.00008-5.

Hartinger, D., & Moll, W. (2011). Fumonisin elimination and prospects for detoxification by enzymatic transformation. *World Mycotoxin Journal*, *4*(3), 271–283. https://doi.org/10.3920/WMJ2011.1285.

Heinl, S., Hartinger, D., Thamhesl, M., Schatzmayr, G., Moll, W. D., & Grabherr, R. (2011). An aminotransferase from bacterium ATCC 55552 deaminates hydrolyzed fumonisin B_1. *Biodegradation*, *22*(1), 25–30. https://doi.org/10.1007/s10532-010-9371-y.

Heinl, S., Hartinger, D., Thamhesl, M., Vekiru, E., Krska, R., Schatzmayr, G., Moll, W. D., & Grabherr, R. (2010). Degradation of fumonisin B_1 by the consecutive action of two bacterial enzymes. *Journal of Biotechnology*, *145*(2), 120–129. https://doi.org/10.1016/j.jbiotec.2009.11.004.

International Agency for Research on Cancer (IARC). (2002). Monographs on the evaluation of carcinogenic risks to humans. In Some traditional herbal medicines, some mycotoxins, naphthalene and styrene, Vol. 82, pp. 1–556. Lyon, France.

Karlovsky, P., Suman, M., Berthiller, F., De Meester, J., Eisenbrand, G., Perrin, I., Oswald, I. P., Speijers, G., Chiodini, A., Recker, T., & Dussort, P. (2016). Impact of food processing and detoxification treatments on mycotoxin contamination. *Mycotoxin Research*, *32*(4), 179–205. https://doi.org/10.1007/s12550-016-0257-7.

Khatibi, P. A., McMaster, N. J., Musser, R., & Schmale, D. G. (2014). Survey of mycotoxins in corn distillers' dried grains with solubles from seventy-eight ethanol plants in twelve states in the U.S. in 2011. *Toxins*, *6*(4), 1155–1168. https://doi.org/10.3390/toxins6041155.

Knutsen, H. K., Alexander, J., Barregård, L., Bignami, M., Brüschweiler, B., Ceccatelli, S., Cottrill, B., Dinovi, M., Edler, L., Grasl-Kraupp, B., Hogstrand, C., Hoogenboom, L. (Ron), Nebbia, C. S., Petersen, A., Rose, M., Roudot, AC, Schwerdtle, T., Vleminckx, C., Vollmer, G., Wallace, H., Dall'Asta, C., Eriksen, G-S, Taranu, I., Altieri, A., Roldán-Torres, R., & Oswald, I. P. (2018). Risks for animal health related to the presence of fumonisins, their modified forms and hidden forms in feed. *EFSA Journal*, *16*(5), e05242. https://doi.org/10.2903/j.efsa.2018.5242.

Kolawole, O., De Ruyck, K., Greer, B., Meneely, J., Doohan, F., Danaher, M., & Elliott, C. (2021). Agronomic factors influencing the scale of *Fusarium* mycotoxin contamination of oats. *Journal of Fungi*, *7*(11), 965. https://doi.org/10.3390/JOF7110965.

Kumar, V., Bahuguna, A., Ramalingam, S., Dhakal, G., Shim, J. J., & Kim, M. (2021). Recent technological advances in mechanism, toxicity, and food perspectives of enzyme-mediated aflatoxin degradation. *Critical Reviews in Food Science and Nutrition*, *62*(20), 5395–5412. https://doi.org/10.1080/10408398.2021.2010647.

Li, C. H., Li, W. Y., Hsu, I. N., Liao, Y. Y., Yang, C. Y., Taylor, M. C., Liu, Y. F., Huang, W. H., Chang, H. H., Huang, H. Lo, Lo, S. C., Lin, T. Y., Sun, W. C., Chuang, Y. Y., Yang, Y. C., Fu, R. H., & Tsai, R. T. (2019). Recombinant Aflatoxin-Degrading F420H2-Dependent reductase from *Mycobacterium smegmatis* protects mammalian cells from aflatoxin toxicity. *Toxins*, *11*(5), 259. https://doi.org/10.3390/toxins11050259.

Li, T., Li, J., Wang, J., Xue, K. S., Su, X., Qu, H., Duan, X., & Jiang, Y. (2024). The occurrence and management of fumonisin contamination across the food production and supply chains. *Journal of Advanced Research*, *60*, 13–26. https://doi.org/10.1016/j.jare.2023.08.001.

Liu, D. L., Yao, D. S., Liang, R., Ma, L., Cheng, W. Q., & Gu, L. Q. (1998). Detoxification of aflatoxin B_1 by enzymes isolated from *Armillariella tabescens*. *Food and Chemical Toxicology*, *36*(7), 563–574. https://doi.org/10.1016/S0278-6915(98)00017-9.

Liu, K. (2011). Chemical composition of distillers grains, a review. *Journal of Agricultural and Food Chemistry*, *59*(5), 1508–1526. https://doi.org/10.1021/jf103512z.

Logrieco, A., Battilani, P., Leggieri, M. C., Jiang, Y., Haesaert, G., Lanubile, A., Mahuku, G., Mesterhazy, A., Ortega-Beltran, A., Pasti, M., Smeu, I., Torres, A., Xu, J., & Munkvold, G. (2021). Perspectives on global mycotoxin issues and management from the mycokey maize working group. *Plant Disease*, *105*(3), 525–537. https://doi.org/10.1094/PDIS-06-20-1322-FE.

Loi, M., De Leonardis, S., Ciasca, B., Paciolla, C., Mulè, G., & Haidukowski, M. (2023). Aflatoxin B_1 degradation by Ery4 laccase: From *in vitro* to contaminated corn. *Toxins*, *15*(5), 310. https://doi.org/10.3390/toxins15050310.

Loi, M., Fanelli, F., Cimmarusti, M. T., Mirabelli, V., Haidukowski, M., Logrieco, A. F., Caliandro, R., & Mule, G. (2018). *In vitro* single and combined mycotoxins degradation by Ery4 laccase from *Pleurotus eryngii* and redox mediators. *Food Control*, *90*, 401–406. https://doi.org/10.1016/j.foodcont.2018.02.032.

Loi, M., Fanelli, F., Liuzzi, V. C., Logrieco, A. F., & Mulè, G. (2017). Mycotoxin biotransformation by native and commercial enzymes: Present and future perspectives. *Toxins*, *9*(4), 111. https://doi.org/10.3390/toxins9040111.

Loi, M., Fanelli, F., Zucca, P., Liuzzi, V. C., Quintieri, L., Cimmarusti, M. T., Monaci, L., Haidukowski, M., Logrieco, A. F., Sanjust, E., & Mulè, G. (2016). Aflatoxin B_1 and M_1 degradation by Lac2 from *Pleurotus pulmonarius* and redox mediators. *Toxins*, *8*(9), 245. https://doi.org/10.3390/toxins8090245.

Loy, D. D., & Lundy, E. L. (2019). Nutritional properties and feeding value of corn and its coproducts. In Sergio O. Serna-Saldivar (Ed.), *Corn: Chemistry and Technology* (3rd ed., pp. 633–659). AACC International by Elsevier Inc. https://doi.org/10.1016/B978-0-12-811971-6.00023-1.

Magan, N., & Aldred, D. (2007). Post-harvest control strategies: minimizing mycotoxins in the food chain. *International Journal of Food Microbiology*, *119*(1-2), 131–139. https://doi.org/10.1016/j.ijfoodmicro.2007.07.034.

Magnoli, A. P., Poloni, V. L., & Cavaglieri, L. (2019). Impact of mycotoxin contamination in the animal feed industry. *Current Opinion in Food Science*, *29*, 99–108. https://doi.org/10.1016/j.cofs.2019.08.009.

Mallmann, C. A., Simões, C. T., Kobs Vidal, J., Rosa Da Silva, C., De Lima Schlösser, L. M., & Araújo De Almeida, C. A. (2021). Occurrence and concentration of mycotoxins in maize dried distillers' grains produced in Brazil. *World Mycotoxin Journal*, *14*(3), 259–268. https://doi.org/10.3920/WMJ2020.2669.

Mani, P., Kumar, V. T. F., Keshavarz, T., Sainathan Chandra, T., & Kyazze, G. (2018). The role of natural laccase redox mediators in simultaneous dye decolorization and power production in microbial fuel cells. *Energies*, *11*(12), 3455. https://doi.org/10.3390/en11123455.

Martínez Tuppia, C., Atanasova-Penichon, V., Chéreau, S., Ferrer, N., Marchegay, G., Savoie, J. M., & Richard-Forget, F. (2017). Yeast and bacteria from ensiled high moisture maize grains as potential mitigation agents of fumonisin B_1. *Journal of the Science of Food and Agriculture*, *97*(8), 2443–2452. https://doi.org/10.1002/jsfa.8058.

Martínez, A., Raybaudi, R., Alvarado, R., Gomez, N., Escobar, E., & Casanova, R. (2004). Occurrence of aflatoxins and fumonisins in corn and rice from Venezuela and its mycoflora. In *XI International IUPAC Symposium on Mycotoxins and Phycotoxins*, Bethesda, MD, USA.

Mendoza, J. R., Rodas, A., Oliva, A., Sabillón, L., Colmenares, A., Clarke, J., Hallen-Adams, H. E., Campabadal, C., & Bianchini, A. (2018). Safety and quality assessment of smallholder farmers' maize in the western highlands of Guatemala. *Journal of Food Protection*, *81*(5), 776–784. https://doi.org/10.4315/0362-028X.JFP-17-355.

Mercosur (2002). MERCOSUR/GMC/Res. N° 25/02. Reglamento técnico MERCOSUR sobre límites máximos de aflatoxinas admisibles en leche, maní y maíz. Derogación de la Res. GMC Nº 56/94.

Ministerio de Economía de la República Argentina (MERA). (2025a, February 20). *Estimaciones agrícolas, Informe Mensual*. Secretaría de Agricultura, Ganadería y Pesca. https://www.magyp.gob.ar/sitio/areas/estimaciones/_archivos/estimaciones/250000_2025/250200_Febrero/250220_Informe%20mensual%20al%2020_02_2025.pdf.

Ministerio de Economía de la República Argentina (MERA). (2025b, Enero). *Informe de Bioenergía*. Secretaría de Agricultura, Ganadería y Pesca, Coordinación de Bioenergía. https://www.magyp.gob.ar/sitio/areas/bioenergia/informes/_archivos//000009_Informes%20Bioenerg%C3%ADa%202025/250100_Informe%20Bioenerg%C3%ADa%20(Enero%202025).pdf.

Mohammadi Shad, Z., Venkitasamy, C., & Wen, Z. (2021). Corn distillers dried grains with solubles: Production, properties, and potential uses. *Cereal Chemistry*, *98*(5), 999–1019. https://doi.org/10.1002/cche.10445.

Morris Navarro, L. F. (2011). Determinación de aflatoxinas en muestras de maíz (Zea mays) y arroz (Oryza sativa) para consumo humano en cinco departamentos de la Costa Caribe Colombiana mediante cromatografía de alta eficiencia durante seis meses en 2011 [Doctoral dissertation, Universidad Nacional de Colombia]. https://repositorio.unal.edu.co/bitstream/handle/unal/8318/598921.2011.pdf?sequence=1&isAllowed=y.

Moschini, R., Borsarelli, M., Martinez, M., Presello, D., Ferraguti, F., Cristos, D., & Rojas, D. (2020). Analysis of preharvest meteorological conditions in relation to concentration of fumonisins in kernels of two genetically different maize hybrids. *Australasian Plant Pathology*, *49*(6), 665–677. https://doi.org/10.1007/s13313-020-00742-4.

Munkvold, G. P., Arias, S., Taschl, I., & Gruber-Dorninger, C. (2019). Mycotoxins in corn: Occurrence, impacts, and management. In S. O. Serna-Saldivar (Ed.), *Corn: Chemistry and Technology* (3rd ed., pp. 235–287). AACC International by Elsevier Inc. https://doi.org/10.1016/B978-0-12-811971-6.00009-7.

Nada, S., Nikola, T., Bozidar, U., Ilija, D., & Andreja, R. (2022). Prevention and practical strategies to control mycotoxins in the wheat and maize chain. *Food Control*, *136*, Article 108855. https://doi.org/10.1016/j.foodcont.2022.108855.

Nguyen, T., Flint, S., & Palmer, J. (2020). Control of aflatoxin M_1 in milk by novel methods: A review. *Food Chemistry*, *311*, Article 125984. https://doi.org/10.1016/j.foodchem.2019.125984.

Nunes, V. M., Moosavi, M., Mousavi Khaneghah, A., & Oliveira, C. A. (2021). Innovative modifications in food processing to reduce the levels of mycotoxins. *Current Opinion in Food Science*, *38*, 155–161. https://doi.org/10.1016/j.cofs.2020.11.010.

Oliveira, M. S., Rocha, A., Sulyok, M., Krska, R., & Mallmann, C. A. (2017). Natural mycotoxin contamination of maize (*Zea mays* L.) in the South region of Brazil. *Food Control*, *73*, 127–132. https://doi.org/10.1016/j.foodcont.2016.07.033.

Pacin, A. M., Bovier, E. C., González, H. H. L., Whitechurch, E. M., Martínez, E. J., & Resnik, S. L. (2009). Fungal and fumonisins contamination in Argentine maize (*Zea mays* L.) silo bags. *Journal of Agricultural and Food Chemistry*, *57*(7), 2778–2781. https://doi.org/10.1021/jf803609c.

Peivasteh-Roudsari, L., Pirhadi, M., Shahbazi, R., Eghbaljoo-Gharehgheshlaghi, H., Sepahi, M., Mirza Alizadeh, A., Tajdar-oranj, B., & Jazaeri, S. (2022). Mycotoxins: Impact on health and strategies for prevention and detoxification in the food chain. *Food Reviews International*, *38*(S1), 193–224. https://doi.org/10.1080/87559129.2020.1858858.

Peña Betancourt, S. D., & Posadas Peña, S. D. (2019). Aflatoxins, zearalenone, fumonisins and deoxynivalenol multiple contamination in raw maize (*Zea mays* L.), for human consumption. *EC Nutrition*, *14*(12), 1–5.

Pinotti, L., Ottoboni, M., Giromini, C., Dell'Orto, V., & Cheli, F. (2016). Mycotoxin contamination in the EU feed supply chain: A focus on cereal byproducts. *Toxins*, *8*(2), 45. https://doi.org/10.3390/toxins8020045.

Pok, P. S., García Londoño, V. A., Aransibia, S. P., Vicente, S., Pacín, A. M., & Resnik, S. L. (2020). Free and hidden fumonisins in Argentinean raw maize samples. *World Mycotoxin Journal*, *13*(1), 109–116. https://doi.org/10.3920/WMJ2019.2484.

Ponce-García, N., Palacios-Rojas, N., Serna-Saldivar, S. O., & García-Lara, S. (2021). Aflatoxin contamination in maize: occurrence and health implications in Latin America. *World Mycotoxin Journal*, *14*(3), 247–258. https://doi.org/10.3920/WMJ2020.2666.

Qi, T. F., Renaud, J. B., McDowell, T., Seifert, K. A., Yeung, K. K. C., & Sumarah, M. W. (2016). Diversity of mycotoxin-producing black Aspergilli in Canadian vineyards. *Journal of Agricultural and Food Chemistry*, *64*(7), 1583–1589. https://doi.org/10.1021/acs.jafc.5b05584.

Qu, L., Wang, L., Ji, H., Fang, Y., Lei, P., Zhang, X., Jin, L., Sun, D., & Dong, H. (2022). Toxic mechanism and biological detoxification of fumonisins. *Toxins*, *14*(3), 182. https://doi.org/10.3390/toxins14030182.

Queiroz, V. A. V., De Oliveira Alves, G. L., Da Conceição, R. R. P., Guimarães, L. J. M., Mendes, S. M., De Aquino Ribeiro, P. E., & Da Costa, R. V. (2012). Occurrence of fumonisins and zearalenone in maize stored in family farm in Minas Gerais, Brazil. *Food Control*, *28*(1), 83–86. https://doi.org/10.1016/j.foodcont.2012.04.039.

Renewable Fuels Association (RFA). (2025). Annual Ethanol Production. https://ethanolrfa.org/markets-and-statistics/annual-ethanol-production.

Reyes-Velázquez, W. P., Espinoza, V. H. I., Rojo, F., Jiménez-Plasencia, C., de Lucas Palacios, E., Hernández-Góbora, J., & Ramírez-Álvarez, A. (2008). Occurrence of fungi and mycotoxins in corn silage, Jalisco State, Mexico. *Revista Iberoamerica de Micología*, *25*, 182–185.

Risa, A., Krifaton, C., Kukolya, J., Kriszt, B., Cserháti, M., & Táncsics, A. (2018). Aflatoxin B_1 and zearalenone-detoxifying profile of *Rhodococcus* type strains. *Current Microbiology*, *75*(7), 907–917. https://doi.org/10.1007/s00284-018-1465-5.

Rodrigues, I., & Chin, L. J. (2012). A comprehensive survey on the occurrence of mycotoxins in maize dried distillers' grain and solubles sourced worldwide. *World Mycotoxin Journal*, *5*(1), 83–88. https://doi.org/10.3920/WMJ2011.1308.

Rodríguez-Couto, S. (2019). Fungal laccase: a versatile enzyme for biotechnological applications. In A. N. Yadav, S. Mishra, S. Singh, A. Gupta (Eds.), *Recent*

Advancement in White Biotechnology Through Fungi (Volume 1, pp.429-457), Fungal Biology. Springer, Cham. https://doi.org/10.1007/978-3-030-10480-1_13.

Roucou, A., Bergez, C., Méléard, B., & Orlando, B. (2022). An Agro-Climatic approach to developing a national prevention tool for deoxynivalenol in French maize-growing areas. *Toxins*, *14*(2), 74. https://doi.org/10.3390/TOXINS14020074.

Samuel, M. S., Sivaramakrishna, A., & Mehta, A. (2014). Degradation and detoxification of aflatoxin B_1 by Pseudomonas putida. *International Biodeterioration & Biodegradation*, *86*, 202–209. https://doi.org/10.1016/j.ibiod.2013.08.026.

Sandlin, N., Kish, D. R., Kim, J., Zaccaria, M., & Momeni, B. (2022). Current and emerging tools of computational biology to improve the detoxification of mycotoxins. *Applied and Environmental Microbiology*, *88*(3). https://doi.org/10.1128/aem.02102-21.

Sangare, L., Zhao, Y., Folly, Y. M. innie E., Chang, J., Li, J., Selvaraj, J. N. imal, Xing, F., Zhou, L., Wang, Y., & Liu, Y. (2014). Aflatoxin B_1 degradation by a *Pseudomonas* strain. *Toxins*, *6*(10), 3028–3040. https://doi.org/10.3390/toxins6103028.

Schwartzbord, J. R., & Brown, D. L. (2015). Aflatoxin contamination in Haitian peanut products and maize and the safety of oil processed from contaminated peanuts. *Food Control*, *56*, 114–118. https://doi.org/10.1016/j.foodcont.2015.03.014.

Scussel, V. M., Savi, G. D., Costas, L. L. F., Xavier, J. J. M., Manfio, D., Bittencourt, K. O., Aguiar, K., & Stein, S. M. (2014). Fumonisins in corn (*Zea mays* L.) from Southern Brazil. *Food Additives & Contaminants: Part B*, *7*(2), 151–155. https://doi.org/10.1080/19393210.2013.862745.

Shi, H., Chang, G., Zhang, Y., Zhao, Y., Wang, H., Zhang, J., & Zhu, J. (2024). Biodegradation characteristics and mechanism of aflatoxin B_1 by *Bacillus amyloliquefaciens* from enzymatic and multiomics perspectives. *Journal of Agricultural and Food Chemistry*, *72*(28), 15841–15853. https://doi.org/10.1021/acs.jafc.4c04055.

Silva, F. G., Zanin, L. M. M., Shimizu, C. F., Lopes, D. D., Ribeiro, J. C., Ishikawa, A. T., Itano, E. N., Kawamura, O., & Hirooka, E. Y. (2022). Aflatoxin B_1 in the egg chain: monitoring with specific indirect competitive ELISA in northern Paraná, Brazil. *World Mycotoxin Journal*, *15*(4), 417–428. https://doi.org/10.3920/WMJ2021.2728.

Song, J., Zhang, S., Xie, Y., & Li, Q. (2019). Purification and characteristics of an aflatoxin B_1 degradation enzyme isolated from *Pseudomonas aeruginosa*. *FEMS Microbiology Letters*, *366*(5). https://doi.org/10.1093/femsle/fnz034.

Stracquadanio, C., Luz, C., La Spada, F., Meca, G., & Cacciola, S. O. (2021). Inhibition of mycotoxigenic fungi in different vegetable matrices by extracts of *Trichoderma* species. *Journal of Fungi*, *7*(6), 445. https://doi.org/10.3390/jof7060445.

Stumpf, R., dos Santos, J., Gomes, L. B., Silva, C. N., Tessmann, D. J., Ferreira, F. D., Machinski, M., & Del Ponte, E. M. (2013). *Fusarium* species and fumonisins associated with maize kernels produced in Rio Grande do Sul State for the 2008/09 and 2009/10 growing seasons. *Brazilian Journal of Microbiology*, *44*(1), 89–95. https://doi.org/10.1590/S1517-83822013000100012.

Sun, X., Sun, C., Zhang, X., Zhang, H., Ji, J., Liu, Y., & Tang, L. (2016). Aflatoxin B_1 decontamination by UV-mutated live and immobilized *Aspergillus niger*. *Food Control*, *61*, 235–242. https://doi.org/10.1016/j.foodcont.2015.09.017.

Torres, O., Matute, J., Gelineau-Van Waes, J., Maddox, J. R., Gregory, S. G., Ashley-Koch, A. E., Showker, J. L., Voss, K. A., & Riley, R. T. (2015). Human health implications from co-exposure to aflatoxins and fumonisins in maize-based foods in Latin America: Guatemala as a case study. *World Mycotoxin Journal*, *8*(2), 143–160. https://doi.org/10.3920/WMJ2014.1736.

United States Department of Agriculture (USDA). (2024a, April 17). *Grain and Feed Annual*. Foreign Agricultural Service. https://apps.fas.usda.gov/newgainapi/api/Report/DownloadReportByFileName?fileName=Grain%20and%20Feed%20Annual_Buenos%20Aires_Argentina_AR2024-0006.pdf.

United States Department of Agriculture (USDA). (2024b, December 17). *Grain and Feed Update*. Foreign Agricultural Service. https://apps.fas.usda.gov/newgainapi/api/Report/DownloadReportByFileName?fileName=Grain%20and%20Feed%20Update_Brasilia_Brazil_BR2024-0038.pdf.

United States Department of Agriculture (USDA). (2024c, August 31). *Biofuels Annual*. Foreign Agricultural Service. https://apps.fas.usda.gov/newgainapi/api/Report/DownloadReportByFileName?fileName=Biofuels%20Annual_Brasilia_Brazil_BR2024-0022.pdf.

United States Department of Agriculture (USDA). (2025a, February). *World Agricultural Production*. Foreign Agricultural Service. https://downloads.usda.library.cornell.edu/usda-esmis/files/5q47rn72z/8049j282k/6h442p53q/production.pdf.

United States Department of Agriculture (USDA). (2025b, February). *Grain: World Markets and Trade*. Foreign Agricultural Service. https://www.fas.usda.gov/sites/default/files/2025-02/grain_0.pdf.

United States Department of Agriculture (USDA). (2025c, January 24). *Grain and Feed Update*. Foreign Agricultural Service https://apps.fas.usda.gov/newgainapi/api/Report/DownloadReportByFileName?fileName=Grain%20and%20Feed%20Update_Buenos%20Aires_Argentina_AR2025-0001.pdf.

Urugo, M. M., Teka, T. A., Berihune, R. A., Teferi, S. L., Garbaba, C. A., Adebo, J. A., Woldemariam, H. W., & Astatkie, T. (2023). Novel non-thermal food processing techniques and their mechanism of action in mycotoxins decontamination of foods. *Innovative Food Science & Emerging Technologies*, *85*, Article 103312. https://doi.org/10.1016/j.ifset.2023.103312.

Vega, L. P., Bautista, K. T., Campos, H., Daza, S., & Vargas, G. (2024). Biofuel production in Latin America: A review for Argentina, Brazil, Mexico, Chile, Costa Rica and Colombia. *Energy Reports*, *11*, 28–38. https://doi.org/10.1016/j.egyr.2023.10.060.

Viaro, H. P., da Silva, J. J., de Souza Ferranti, L., Bordini, J. G., Massi, F. P., & Fungaro, M. H. P. (2017). The first report of *A. novoparasiticus*, *A. arachidicola* and *A. pseudocaelatus* in Brazilian corn kernels. *International Journal of Food Microbiology*, *243*, 46–51. https://doi.org/10.1016/j.ijfoodmicro.2016.12.002.

Voss, K. A., Smith, G. W., & Haschek, W. M. (2007). Fumonisins: Toxicokinetics, mechanism of action and toxicity. *Animal Feed Science and Technology*, *137*(3–4), 299–325. https://doi.org/10.1016/j.anifeedsci.2007.06.007.

Wang, J., & Xie, Y. (2020). Review on microbial degradation of zearalenone and aflatoxins. *Grain & Oil Science and Technology*, *3*(3), 117–125. https://doi.org/10.1016/j.gaost.2020.05.002.

Wang, X., Qin, X., Hao, Z., Luo, H., Yao, B., & Su, X. (2019). Degradation of four major mycotoxins by eight manganese peroxidases in presence of a dicarboxylic acid. *Toxins*, *11*(10), 566. https://doi.org/10.3390/toxins11100566.

Wigmann, É. F., Meyer, K., Cendoya, E., Maul, R., Vogel, R. F., & Niessen, L. (2020). A loop-mediated isothermal amplification (LAMP) based assay for the rapid and sensitive group-specific detection of fumonisin producing *Fusarium* spp. *International Journal of Food Microbiology*, *325*, 108627. https://doi.org/10.1016/j.ijfoodmicro.2020.108627.

World Health Organization (WHO) (2023, October 2). *Mycotoxins*. https://www.who.int/es/news-room/fact-sheets/detail/mycotoxins.

Wu, Y. Z., Lu, F. P., Jian, H. L., Tan, C. P., Yao, D. S., Xie, C. F., & Liu, D. L. (2015). The furofuran-ring selectivity, hydrogen peroxide-production and low Km value are the three elements for highly effective detoxification of aflatoxin oxidase. *Food and Chemical Toxicology*, *76*, 125–131. https://doi.org/10.1016/j.fct.2014.12.004.

Xie, Y., Wang, W., & Zhang, S. (2019). Purification and identification of an aflatoxin B_1 degradation enzyme from *Pantoea* sp. T6. *Toxicon*, *157*, 35–42. https://doi.org/10.1016/j.toxicon.2018.11.290.

Xu, Y., Dong, H., Liu, C., Lou, H., & Zhao, R. (2023). Efficient aflatoxin B_1 degradation by a novel isolate, *Pseudomonas aeruginosa* M-4. *Food Control*, *149*, Article 109679. https://doi.org/10.1016/j.foodcont.2023.109679.

Yehia, R. S. (2014). Aflatoxin detoxification by manganese peroxidase purified from *Pleurotus ostreatus*. *Brazilian Journal of Microbiology*, *45*(1), 127–134. https://doi.org/10.1590/S1517-83822014005000026.

Zeinvand-Lorestani, H., Sabzevari, O., Setayesh, N., Amini, M., Nili-Ahmadabadi, A., & Faramarzi, M. A. (2015). Comparative study of *in vitro* prooxidative properties and genotoxicity induced by aflatoxin B_1 and its laccase-mediated detoxification products. *Chemosphere*, *135*, 1–6. https://doi.org/10.1016/j.chemosphere.2015.03.036.

Zhang, W., Xue, B., Li, M., Mu, Y., Chen, Z., Li, J., & Shan, A. (2014). Screening a strain of *Aspergillus niger* and optimization of fermentation conditions for degradation of aflatoxin B_1. *Toxins*, *6*(11), 3157–3172. https://doi.org/10.3390/toxins6113157.

Zhang, Y., & Caupert, J. (2012). Survey of mycotoxins in U.S. distiller's dried grains with solubles from 2009 to 2011. *Journal of Agricultural and Food Chemistry*, *60*(2), 539–543. https://doi.org/10.1021/jf203429f.

Zhang, Y., Caupert, J., Imerman, P. M., Richard, J. L., & Shurson, G. C. (2009). The occurrence and concentration of mycotoxins in U.S. distillers dried grains with solubles. *Journal of Agricultural and Food Chemistry*, *57*(20), 9828–9837. https://doi.org/10.1021/jf901186r.

Zhao, Z., Zhang, Y., Gong, A., Liu, N., Chen, S., Zhao, X., Li, X., Chen, L., Zhou, C., & Wang, J. (2019). Biodegradation of mycotoxin fumonisin B_1 by a novel bacterial consortium SAAS79. *Applied Microbiology and Biotechnology*, *103*(17), 7129–7140. https://doi.org/10.1007/s00253-019-09979-6.

Zheng, Y., Wu, W., Sun, C., Liu, H., & Dou, J. (2024). Occurrence and fate analysis of mycotoxins in maize during the post-harvest period. *Toxins*, *16*(11), 459. https://doi.org/10.3390/toxins16110459.

Chapter 8

Myco-Synthesized Zinc Oxide Nanoparticles as a Sustainable Tool for Controlling *Aspergillus* Species and Mycotoxin Production in Agriculture

Gabriela A. Pena*

Department of Microbiology and Immunolcgy, National University of Río Cuarto (UNRC), Research Institute of Mycology and Mycotoxicology (IMICO)-Global Research Council (CONICET), Río Cuarto, Córdoba, Argentina, Río Cuarto, Córdoba, Argentina

Abstract

One promising eco-friendly, cost-effective, and safe alternative is the use of biogenic nanoparticles (NPs), particularly zinc oxide nanoparticles synthesized by fungi (Myco-synthesized ZnO-NPs). ZnO nanoparticles have demonstrated antifungal properties, making them a potential tool for controlling fungal growth and mycotoxin production. The 'One Health' approach emphasizes the interconnection between human, animal, and environmental health, which is crucial for managing diseases affecting agriculture and public health.

This approach is particularly relevant in the agro-industrial sector, which faces the challenge of feeding a growing global population projected to reach around 10.3 billion people in the mid-2080s. The sector must balance increasing food production with ensuring food safety and quality. Additionally, climate change is expected to impact the development of pathogens, changing the susceptibility of hosts and creating new conditions for diseases. Fungal diseases are a major threat to many crops.

* Corresponding Author's Email: gpena@exa.unrc.edu.ar

In: Biocontrol Agents (BCAs)
Editor: Daiana García
ISBN: 979-8-90134-102-5

In Argentina, corn (*Zea mays* L.) is a major crop, as the country is the fifth-largest producer and fourth-largest exporter globally.
However, fungal contamination can lead to the production of mycotoxins, which contaminate crops under several environmental conditions. Among the *Aspergillus* species, *A. flavus* is particularly problematic, producing aflatoxins (AFs), which are highly toxic, carcinogenic, and genotoxic. These toxins pose significant health risks to both animals and humans, especially in the context of food consumption. This review provides an overview of conventional chemical and physical methods for synthesizing ZnO-NPs. Special emphasis is placed on recent advances in using fungal strains, particularly filamentous fungi, for their myco-synthesis. It also discusses the conditions influencing their production, their demonstrated antifungal efficacy, and explores their potential application in controlling toxigenic fungi and their associated mycotoxins. While numerous studies have successfully achieved the myco-synthesis of ZnO-NPs and confirmed their antifungal properties, primarily in vitro, no research has yet investigated their potential to reduce mycotoxin production in crops. Further studies are also required to evaluate the environmental impact of myco-synthesized ZnO-NPs through toxicological assessments to confirm their low or negligible toxicity.

Keywords: ZnO-NPs, myco-synthesis, myco-synthesized ZnO-NPs, *Aspergillus*, mycotoxins, Aflatoxins

Introduction

Nanocompounds are widely used in many fields such as environmental science, medicine, and agriculture. NPs are particles with at least one dimension less than 100 nm, such as nanosheets, nanotubes, and nanowires. In addition to their unique physical and chemical properties, NPs act as a bridge between bulk materials and atomic or molecular structures [1]. Indeed, compared to their conventional bulk counterparts, NPs possess completely different or enhanced characteristics such as decreased size, enhanced specific surface area, and thus strengthened reactivity [2]. Nanotechnology has become a promising option for controlling crop diseases caused by microorganisms, with growing applications in food production and preservation [3, 4], as well as in the development of nanopesticides and nanofertilizers [5-6]. Nanotechnology offers multiple applications in agriculture, including crop growth, production, storage, packaging, and transport. Its main advantage over conventional methods lies in enhancing crop resilience to environmental stresses, such as drought, salinity, and elevated CO_2, thereby improving productivity. Nanomaterials are also used in biosensors, genetic transformation, and nano-based growth regulators [7]. Metallic nanoparticles are well known for their efficient adsorptive capacity and catalytic potential in degrading dye-contaminated water. Additionally, various metal-based nanoparticles such as silver, copper, zinc, nickel, and titanium, among others, have demonstrated notable antifungal properties [8]. Among other metal nanoparticles, ZnO-NPs have considerable advantages due to their low cost, strong antibacterial and antifungal activities [9-13] as well as being an essential micronutrient in agriculture [14]. In addition to their known antibacterial and antifungal effects, ZnO-NPs can act as antioxidant and antidiabetic agents [15].

Nowadays, these nanocompounds are mainly synthesized by chemical methods, causing environmental pollution and posing potential health risks. The biogenic synthesis of NPs represents an attractive nanotechnological alternative to conventional chemical and physical approaches, offering several advantages: reduced energy and resource consumption, relatively low cost, lower generation of toxic waste, improved environmental performance, and higher synthesis yields [16-17].

Biological sources such as plant extracts, algae, yeast, bacteria, and fungi can act as reducing and stabilizing agents in the synthesis of functional NPs. Among these, synthesis mediated by fungi offers distinct advantages over plants due to their ability to produce large quantities of secondary metabolites and bioactive compounds, such as enzymes, which promote the high-yield production of stable nanoparticles [16]. This fungus-mediated synthesis, known as *myco-synthesis*, has gained increasing attention for its potential applications in sustainable agriculture and as an antimicrobial strategy effective against both plant and human pathogens. The "*One Health*" approach emphasizes the interaction between human, animal, and environmental health, which is critical for the prevention and control of diseases affecting plants, animals, and humans. This approach is particularly relevant in the agro-industrial sector, which faces the challenge of feeding a growing global population projected to reach around 10.3 billion people in the mid-2080s [18]. The sector must balance increasing food production with ensuring food safety and quality. Additionally, climate change is expected to impact the development of pathogens, changing the susceptibility of hosts and creating new conditions for diseases. Fungal diseases are a major threat to many crops. In Argentina, corn (*Zea mays* L.) is a key agricultural product. The country ranks as the fifth-largest producer and fourth-largest exporter of maize worldwide [19]. Argentina exports around 70% of the total maize produced, whereas 30% is destined to animal feed and 8% to ethanol production, wet and dry maize grinding, and other industries [20]. However, fungal contamination can lead to the production of mycotoxins, which contaminate crops under several environmental conditions. Mycotoxins are toxic compounds with immunological, teratogenic, and carcinogenic effects in humans and animals, which are not essential to the life cycle of fungi but give them adaptive and competitive advantages.

Toxigenic species of the genus *Aspergillus* are commonly found in extensive crops, especially in maize and peanuts [21]. *Aspergillus flavus* and *A. parasiticus* are the most frequently isolated species that cause crop deterioration and, more importantly, produce AFs. These are a group of natural mycotoxins that are highly toxic, carcinogenic, and genotoxic, and have been classified by the International Agency for Research on Cancer (IARC) as a Group IA carcinogen [22]. AFs are considered the most potent naturally occurring carcinogens and pose a significant threat to animal and human health through food consumption.

Also, as is predicted, with an increasing world population, crop global production needs to increase by up to 119 percent; at the same time, waste and spoilage of foods should be reduced [18]. Therefore, among many strategies to prevent mycotoxin production, one promising, eco-friendly, cost-effective, and safe alternative is the use of biogenic nanoparticles, particularly ZnO-NPs obtained by myco-synthesis, to control toxigenic fungal species and, specifically, the production of AFs in extensive crops.

Zinc is an essential micronutrient for both humans and animals and plays a crucial role in DNA replication and repair, oxidative stress regulation, and cell cycle progression. Compared to conventional sources of Zinc, ZnO-NPs exhibit lower toxicity, higher absorption rates, greater bioavailability, enhanced immune support, antioxidant potential, superior biocompatibility, and potent antimicrobial effects [23-24].

Conventional Physical and Chemical Methods for the Synthesis of ZnO Nanoparticles

Several methods have been developed for the synthesis of ZnO-NPs, including chemical, physical, and biological approaches, to obtain particles with desirable size and morphologies for diverse applications. Generally, two main strategies are used to synthesize ZnO-NPs through physical or chemical methods: the top-down and the bottom-up approaches. The top-down approach involves the fragmentation of bulk materials into nanoscale particles and is typically categorized as a physical method. In contrast, the bottom-up approach builds nanoparticles from smaller units such as atom or molecule precursors and is commonly associated with chemical methods. The following subsection describes some of the most widely used techniques within these two strategies.

Top-down approaches:

- *Combustion method:* is a fast, low-cost, and energy-efficient process, using self-propagating or volume combustion modes. Produces ZnO-NPs (17-74 nm) with high purity from the ignition of various precursors like zinc acetate and urea [25, 26].
- *Thermal decomposition:* is a simple and scalable method involving the breakdown of compounds like zinc oxalate (aqueous solution) or Zn-MOFs at high temperatures (450-550°C). This process generates crystalline ZnO-NPs ranging from 27 to 100 nm without toxic solvents or surfactants. The solution is first heated to 100°C to remove the water, then to 200°C for complete dehydration. Finally, the resulting powder is heated to 450°C for 6 h to achieve complete decomposition [27].
- *Ultrasonication:* uses high-frequency sound waves to create cavitation, reducing particle size and preventing aggregation. Efficiently produces very small ZnO-NPs (<10 nm), including quantum dots (~2.4 nm).
- *Microwave-assisted combustion:* a rapid method using microwave radiation to heat reactants. Produces ZnO-NPs (2.6-90 nm) in minutes with good crystallinity and controlled morphology.

Other top-down methods for ZnO-NPs synthesis include laser ablation, mechanical grinding, and a one-step solid-phase method [28].

Bottom-up approaches:

- *Co-precipitation* is a simple and cost-effective technique involving the precipitation of metal salts by alkalis. After drying and optional calcination, ZnO-NPs (43-140 nm) are obtained with high yield and purity. Variations of this method have produced ZnO-NPs with different modifiers (e.g., oleic acid, ADPA) and shapes (e.g., spheres, nanosheets).

- *Sol-Gel method:* is a wet chemical technique involving hydrolysis and condensation of metal alkoxides or salts as precursors to form a gel, which is later sintered into ZnO-NPs. It produces highly stable ZnO-NPs ranging from 32 to 200 nm with controlled morphology, including doped variants.
- *Hydrothermal method:* involves synthesizing ZnO nanoparticles in aqueous solutions under high temperature and pressure in autoclaves. This method produces ZnO-NPs in various morphologies (spherical, rod, flower-like) with sizes of 21-500 nm, influenced by temperature, time, and pH.
- *Homogeneous precipitation:* this method is an enhanced form of direct precipitation that ensures uniform distribution of the precipitant by generating it gradually and evenly throughout the solution via controlled chemical reactions. It allows for slow and even release of ions, promoting uniform crystal growth. This method has been used to synthesize hexagonal and rod-like ZnO-NPs without the need for surfactants or gelling agents.
- *Deposition Template Method:* this method involves depositing material onto or into a template, which is later removed to form nanostructures that mimic the template's shape. It is considered a promising approach for biosynthesis and green chemistry. For example, ZnO nanorods (~60 nm) have been synthesized using sodium alginate, and porous ZnO has been obtained using natural wood as a template.

Each method offers different advantages in terms of scalability, cost, complexity, and environmental impact, e.g., top-down methods use costly equipment, a large space area, high temperature, and pressure, but generally are fast, easily scalable, and produce NPs of high quality. Whereas the bottom-up methods are cheap, and relatively large and complex nano-structural systems can be constructed. Therefore, different ZnO-NPs have been created by applying them. As a result, NPs can adopt various nanostructure forms, which depend on the selected synthesis conditions, such as nanorods, nanotubes, nanoflowers, and nanobelts, among others. Further, their capping with stabilizers and biocompatible polymers prevents their aggregation and reduces nonspecific protein adsorption on their surface in biological media, as well as shows lower toxicity [29]. Indeed, the ZnO-NPs surface could be modified by capping using 3-aminopropyltriethoxysilane (APTES) or dopants like metal oxides, carbon derivatives, or N-type semiconductors, improving their dispersion and enhancing their photocatalytic performance [30-31].

NPs Aggregation

Nanoscale particles are inherently unstable and tend to aggregate due to attractive forces such as van der Waals, electrostatic, or magnetic interactions. In the absence of sufficient repulsive forces, nanoparticles may suffer agglomeration, aggregation, or coalescence processes [32]. The adsorption of stabilizing agents onto nucleated nanocrystals reduces surface free energy, thereby decreasing the reactivity of the particles. Aggregation of NPs is typically prevented through electrostatic and steric stabilization mechanisms. Stabilizers create a physical barrier that prevents direct contact between metallic surfaces. Additionally, they can alter the surface charge of the particles and affect their colloidal stability. The combined effects of surface energy reduction by surfactants, charge-charge interactions, and steric repulsion between particles help prevent nanoparticle aggregation [33].

Electrostatic stabilization is achieved through the formation of at least one electrical double layer on the nanoparticle's surface. This mechanism relies on repulsive forces between similarly charged surfaces in solution, which prevent aggregation. It is the primary method used to stabilize nanoparticles in aqueous solutions [33]. Steric stabilization involves the adsorption of large molecules, such as polymers or surfactants, onto the surface of nanoparticles, forming a protective layer that prevents aggregation. This type of stabilization results from steric repulsion between the large hydrophobic chains of the stabilizing agents. In general, steric mechanisms play a dominant role in stabilizing metallic colloids in non-polar solvents [34]. The type of polymer used is often critical when functionalizing nanoparticles. Electrosteric stabilization combines both electrostatic and steric mechanisms. Ionic surfactants commonly used as stabilizing agents contain a polar head group that can generate an electrical double layer and a lipophilic tail that provides steric repulsion. The nature of the stabilizer surrounding the nanoparticle is a key factor in determining its catalytic activity [35], antibacterial properties [36], suitability for medical applications [37], and interaction with the natural environment [38].

Practical Applications of the ZnO-NPs

The ZnO-NPs have received significant attention in recent years due to their unique properties and diverse applications across various research fields, particularly in agriculture and medicine, as well as in the food and cosmetic industries.

One of the main advantages of ZnO-NPs lies in the classification of zinc oxide as GRAS (Generally Recognized as Safe) by the U.S. Food and Drug Administration (FDA). ZnO-NPs, both alone and doped with other metals, appear as a novel candidate for applications such as tissue regeneration, implant coatings, bioimaging, wound healing, cancer therapy development, targeted drug delivery, antimicrobial coatings and bandages, biosensors, and gene delivery systems [39]. Moreover, zinc oxide has also been categorized as a GRAS material by the European Food Safety Authority (EFSA) for use in animal and plant products [40]. Interestingly, in clinical settings, ZnO-NPs have demonstrated specific toxicity toward cancer cells through their dissolution at low pH, which releases Zn^{2+} ions, generating reactive oxygen species (ROS) and inducing cytotoxicity. Furthermore, ZnO-NPs have been employed as efficient carriers for the targeted delivery of anticancer drugs into tumor cells. Photoirradiation with UV light or near-infrared (NIR) laser can further enhance their anticancer activity through a synergistic chemo-photodynamic effect [41].

Currently, there is a growing need to develop novel strategies to control fungal pathogens in order to prevent plant diseases and post-harvest losses of food crops and grains, as well as to minimize the harmful effects of these fungi on humans and livestock [8].

ZnO-NPs for Controlling Fungal Diseases and Mycotoxin Accumulation in Agriculture

Diseases caused by fungi are a major cause of crop and yield losses in agriculture. Fungi are responsible for numerous plant diseases, and their infections can occur in the field and affect the crops. Additionally, fungal and/or mycotoxin contamination can occur during grain storage. Specifically, cereals are susceptible to fungal infection with plant pathogens and/or toxigenic fungi during both pre- and post-harvest stages. Moreover, mycotoxin levels may increase during post-harvest handling and storage.

Humans and animals can consume mycotoxins either through food and feed or mycotoxin accumulation in animal tissues, milk, and eggs, as well as through processed foods, since these metabolites are resistant to conventional industrial processing methods [42].

The most relevant mycotoxins in relation to food and feed safety are primarily produced by members of four genera of filamentous fungi: *Aspergillus*, *Fusarium*, *Penicillium*, and *Alternaria*. These fungi can colonize a wide range of crops and are adapted to diverse environmental conditions such as varying levels of humidity and temperature, as well as different plant matrices [21, 42]. Although more than 400 mycotoxins have been identified, the most important in relation to their toxicity and health impacts on both humans and animals are AFs, ochratoxin A (OTA), *Fusarium* toxins, such as fumonisins (FBs), zearalenone (ZEA), trichothecenes (TCT), and deoxynivalenol (DON). These compounds pose significant health risks and lead to considerable economic losses. Therefore, taking into account this problem and since their unavoidable presence in cereals, cereal-based products, dairy products, spices, nuts, coffee, vegetable oils, wine, and fruit juices, regulatory limits for maximum levels of AFs, TCT, and FBs have been established by various regulating entities around the world [21, 42-45]. In response to this issue, many studies have investigated the *in vitro* antibacterial and antifungal activities of ZnO-NPs of various sizes and shapes synthesized through different approaches, demonstrating their strong antimicrobial potential. However, research specifically focused on the antifungal effects of ZnO-NPs against toxigenic fungi and their influence on mycotoxin accumulation in agricultural settings remains limited [8].

ZnO-NPs can easily be synthesized from a low-cost precursor such as zinc acetate dihydrate, by reacting it with an alkaline NaOH solution and allowing the dropwise precipitation of ZnO, as proposed by Sharma et al. [46] and applied by Pena et al. [12] without any further treatment of the resulting ZnO-NPs. These ZnO-NPs have been chemically characterized and applied to control fungal growth of *A. flavus* and *Fusarium proliferatum, as well as the production of* their respective mycotoxins, AFs and FBs, in both *in vitro* and *in situ* assays, including field-level trials on maize plants. [12, 47, 49].

Aspergillus Flavus and Aspergillus Parasiticus - The Main Aflatoxin Producers

Aspergillus flavus and *A. parasiticus*, two species within the *Aspergillus* section *Flavi*, are the main AF producers [50-51]. Aflatoxins are mycotoxins of great relevance in terms of toxicity and food safety. Among all the AFs described, four types are the most well-known and frequently found in food: Aflatoxins B_1, B_2, G_1, and G_2 (AFB_1, AFB_2, AFG_1, and AFG_2), with AFB_1 being the most important due to its high toxicity and prevalence. AFB_1 is the most toxic natural compound described and has been classified as a Group 1 human carcinogen by the International Agency for Research on Cancer [22]. Exposure to AFs can lead to hepatotoxicity, immunotoxicity, and teratogenicity [52]. Considering the risks associated with mycotoxin contamination, many national agencies have established maximum permissible limits for each mycotoxin in a wide variety of food and feed products and sub-products. In Latin America, the maximum amount of AFB_1 allowed in maize is 5 μg/kg, while the limit for total AFs in this cereal is 20 μg/kg [53].

Contamination of food and feed with AFs can result in aflatoxicosis, an intoxication caused by ingestion or inhalation of aflatoxins. There are two main forms: acute and chronic. Acute aflatoxicosis results from the intake of high levels of AFs over a short period, leading to liver damage, liver failure, and even death. Chronic aflatoxicosis stems from long-term exposure to low concentrations of AFs and may result in liver cancer, immune suppression, and growth retardation in children. The incidence of aflatoxicosis varies globally each year and is strongly influenced by climatic conditions [52–54].

Aspergillus flavus is widely distributed in temperate, tropical, and subtropical regions, including various agroecosystems and diverse natural habitats. Consequently, there is a potential risk of AF contamination around the world [55]. In Argentina, AFs are strictly regulated by MERCOSUR and the Argentine Food Code (CAA), due to their well-documented adverse effects on both human and animal health, and because maize is a key export commodity for the country [44, 51]. However, the incidence of aflatoxigenic strains and the occurrence of AFB_1 in maize have only been documented in a few studies conducted more than a decade ago [56-57].

In the context of climate change projected for the coming years, AFs are expected to exert an increasingly negative impact on the food and feed production chain [21]. In this regard, Yu et al. [58]. developed a predictive model for AFs risk in the United States, incorporating both maize growth stages and climatic conditions. Comparing data from 2011-2020 with projections for 2031-2040, they found that more than 89% of counties across 15 states are expected to be affected by high levels of AF contamination. These findings highlight the growing global threat that AFs pose in the coming decades. Therefore, there is an urgent need to implement effective strategies to prevent AF contamination in crops and reduce their entry into the human and animal food chains. In response, several biological and chemical strategies have recently been applied to reduce or inhibit mycotoxin production by *A. flavus* and *A. parasiticus* in Argentina [21, 46, 47, 59].

Myco-Synthesized ZnO-NPs to Control Fungi and Mycotoxins in Agriculture

Efforts are underway to establish effective strategies for controlling fungal growth and the associated mycotoxin production. As a result, there is growing interest in biological approaches and green nanotechnology aiming to minimize the use of chemical agents and promote environmentally friendly methods [60]. "Green nanotechnology" is a term that refers to a method for producing nanomaterials that have low environmental impact or provide ecological benefits by eliminating or minimizing the usage of hazardous materials. Their key principles include the use of water as a solvent, eco-friendly reducing agents, and non-toxic stabilizers, with biological systems offering ideal conditions for such processes [61]. Various organisms, including bacteria, fungi, actinomycetes, algae, and plants, have been used to biosynthesize NPs. This bottom-up process mainly involves redox reactions facilitated by microbial enzymes or plant phytochemicals [62]. Microorganisms are especially attractive due to their simplicity, cost-effectiveness, and ability to convert metal ions into NPs. Additionally, natural biomolecules act as capping agents, stabilizing and functionalizing NPs in an environmentally friendly manner [1, 63].

General Mechanisms of Nanoparticles Myco-Synthesis

The use of fungal cultures or biomass in nanoparticle synthesis, known as *myco-synthesis*, is a simple and environmentally friendly approach that has been explored as an alternative method for producing nanoparticles. The main advantages of using filamentous fungi for the synthesis of metallic NPs include their high tolerance to metals, strong bioaccumulation capacity, ability to secrete a wide range of enzymes, and easy manipulation [64-66]. Compared to bacteria, fungi can synthesize a larger quantity of NPs due to their higher protein secretion, resulting in increased productivity [65]. They can also tolerate or adapt to a wider range of growth conditions compared to other microorganisms and plants, including resistance to agitation, flow pressure, and other stresses within chambers and bioreactors [68-69].

Two mechanisms of fungal NP synthesis have been reported: intracellular and extracellular. Intracellular synthesis is less common and requires several post-synthesis steps for the purification of metal NPs, making it inefficient and cost ineffective. In contrast, extracellular synthesis is more widely used due to its simplicity and economic advantages. The general methodology for the myco-synthesis of metal NPs involves growing the fungus in a nutrient-rich medium for 3–5 days. This culture can be used directly as a source of secondary metabolites and biomolecules, or the fungal biomass can be harvested and incubated in distilled water for 2–4 days. Then, the supernatant containing the fungal molecules is recovered and placed in contact with the different precursors, at different concentrations and incubation times. Thus, fungal metabolites and other biomolecules carry out the reduction of the precursors and, consequently, the precipitation of the metal NPs, generally observed as colored precipitates, depending on the synthesized NP [66]. Figure 1 shows a schematic representation of the intracellular and extracellular mechanisms involved in myco-synthesis.

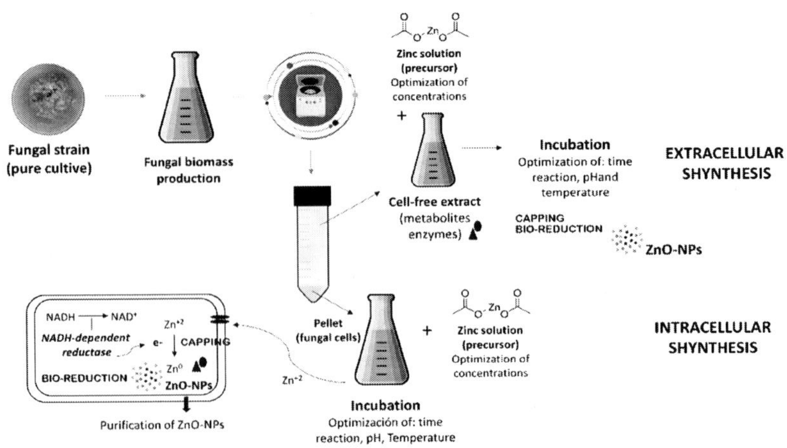

Figure 1. Mechanisms of fungal-mediated synthesis of ZnO-NPs.

Overview of Fungal Species Involved in the Myco-Synthesis of ZnO-NPs and Their Applications in Agriculture

As previously mentioned, the synthesis of ZnO-NPs using filamentous fungi is a simple and straightforward process that allows the production of functional ZnO-NPs through cost-effective, eco-friendly, and efficient methods. However, not all fungi are able to generate ZnO-NPs. Therefore, selecting the appropriate fungal isolate is a crucial preliminary step in the myco-synthesis approach. Some reports suggest that molecules involved in myco-synthesis include enzymes or enzyme extracts, such as xylanases and others. Consequently, it is recommended to use highly enzymatic fungi such as species of the genus *Trichoderma* or fungi from the phylum *Basidiomycetes*.

Numerous studies have reported the myco-synthesis of ZnO-NPs using various fungal strains. Accordingly, some of the most notable findings are outlined below. Additionally, Table 1 presents a non-exhaustive summary of myco-synthesized ZnO-NPs synthesized by different fungal species and their demonstrated applications. Several studies have employed different *Trichoderma* species to produce ZnO-NPs and to evaluate their antifungal effects against soil-borne and crop pathogens. These studies found that myco-synthesized ZnO-NPs, when applied to the plant system in greenhouse assays, not only reduced disease incidence but also improved plant growth parameters [67-68]. Ghareeb et al. [68] using filtrates of *T. longibrachiatum* for producing ZnO-NPs reported a significant inhibitory effect on the growth of *F. oxysporum* both *in vitro* and under greenhouse conditions. Additionally, they observed a significant reduction in the incidence of *Meloidogyne incognita*, a nematode pest affecting sweet pepper. Moreover, the application of the myco-synthesized ZnO-NPs significantly improved plant quality and yield.

Moreover, in addition to *Trichoderma* species, several other *Ascomycetes* have been used for the myco-synthesis of ZnO-NPs. Mekky et al. [69] synthesized ZnO-NPs using extracellular extracts of *A. niger*. The resulting ZnO-NPs exhibited antibacterial activity against *Klebsiella pneumoniae*, *Escherichia coli*, *Acinetobacter baumannii*, *Pseudomonas aeruginosa*, *Staphylococcus aureus,* and *S. haemolyticus*, as well as antifungal activity against *A. niger*, *Penicillium marneffei*, *Candida glabrata,* and *C. parapsilosis*, as demonstrated by agar well diffusion assays. In another study, Tamin et al. [70] used culture filtrates of *A. niger* combined with a zinc carbonate solution.

Table 1. Myco-synthesized ZnO-NPs by using different fungal species and their applications

Fungi	Mechanism of synthesis	Precursor	Average size and shape of ZnO-NPs	Culture conditions*	Applications	References
Trichoderma harzianum	Extracellular using biomass filtrate	Zinc acetate	13 nm, spherical	24-48 h, shaking at 150 rpm, 40°C	Antifungal against *F. oxysporum* (*in vitro*)	Ahmed et al. [66]
Trichoderma harzianum	Extracellular using cell-free filtrate	Zinc acetate dihydrate	8-25 nm, mixture of hexagonal, spherical and rod-shaped NPs, either individual or agglomerated	72 h, shaking at 150 rpm	Antifungal against three soil-borne fungal pathogens of cotton (*in vitro* and greenhouse assay).	Zaki et al. [67]
Trichoderma longibrachiatum	Fungal suspension of spores and metabolites	Zinc sulphate	10.5 nm	5 days, shaking at 150 rpm	Anti-nematode and antifungal against pepper soil-borne pathogens (*in vitro* and greenhouse assay)	Ghareeb et al. [68]
Aspergillus niger	Extracellular extracts	Zinc acetate	20 ± 5 nm, hexagonal crystal structure	72 h, shaking at 150 rpm, pH~6.5, 32°C in darkness	Antibacterial, antifungal and anticancer activities (*in vitro*)	Mekki et al. [69]
Aspergillus niger	Extracellular using cell-free filtrate	Zinc carbonate	27-40 nm	1 h shaking at 70°C pH 10-12. Then, 12 h digestion in mother liquor at 90°C	Antifungal against *A. ochraceus* and *A. niger* (*in vitro*)	Tamin et al. [70]
Xylaria arbuscula	Extracellular using cell-free filtrate	Zinc acetate	21 nm, hexagonal	72 h, shaking at 150 rpm, 28°C, pH 9	Antibacterial, antifungal, dye degradation and wound healing acceleration (*in vitro*)	Nehru et al. [71]

Fungi	Mechanism of synthesis	Precursor	Average size and shape of ZnO-NPs	Culture conditions*	Applications	References
Xylaria acuta	Fungal filtrate, combustion method	Zinc nitrate hexahydrate	34-55 nm, hexagonal	Magnetic stirring, muffling furnace at 400 ± 10°C	Antibacterial (Gram-positive and -negative), antifungal and anti-cancer activities (*in vitro*)	Sumanth et al. [73]
Aspergillus terreus	Extracellular using cell-free filtrate	Zinc acetate	30 nm, spherical with irregular margins	2 h under vigorous stirring at room temperature	Antibacterial, antifungal and antioxidant activities (*in vitro*)	Mousa et al. [74]
Aspergillus fumigatus	Extracellular using cell-free filtrate	Zinc sulphate	60-30 nm, spherical	72 h, shaking at 150 rpm, 32°C, pH 6.5	Antibacterial (Gram-positive and -negative) (*in vitro*)	Rajan et al. [75]
Aspergillus niger	Extracellular using cell-free filtrate	Zinc nitrate	45 nm, spherical	96 h, shaking at 180 rpm, 30°C	Antibacterial against fish pathogens, promoting fish growth	El-Saadony et al. [76]
Pleurotus floridanus	Extracellular using cell-free filtrate	Zinc nitrate	35 nm, spherical	Various culture conditions, optimal: pH 10, 300 mM precursor	Optimization of myco-synthesis by comparing three synthesis methods	Rafeeq et al. [77]
Aspergillus niger	Extracellular using cell-free filtrate	Zinc acetate	82-176 nm, irregular/quaternary shapes	7 days, shaking at 180 rpm, 37°C.	Antibacterial against carbapenem-resistant *K. pneumoniae* (KPC)	Rasha et al. [78]

* All procedures included a final centrifugation step and, generally, a drying step to obtain the ZnO-NPs.

The ZnO-NPs characterization, throughout various methods, including TEM, SEM, and EDX, revealed highly crystalline ZnO-NPs with an average size of 27-40 nm. The antifungal effect of the myco-synthesized ZnO-NPs against *A. ochraceus* and *A. niger* strains showed up to 87% inhibition of fungal growth *in vitro*.

Nehru et al. [71] produce ZnO-NPs using the biomass filtrate of the endophytic fungus *Xylaria arbuscula*, isolated from *Blumea axillaris* Linn, as a capping agent and zinc acetate as the precursor salt. The biosynthesized ZnO-NPs were characterized by spectroscopic (plasmon peak) and microscopic methods (SEM, TEM, EDX, and Z-potential). These nanoparticles showed significant concentration-dependent inhibition of bacterial strains (*S. aureus*, *E. coli*, *Enterococcus faecalis*, *P. aeruginosa*, *Proteus mirabilis*, *S. epidermidis*, *K. pneumoniae,* and *Bacillus subtilis*) and fungal species (*C. albicans* and *A. niger*). Both assays were determined by disc diffusion agar. *In vitro* cytotoxicity and wound healing potential of ZnO-NPs were also evaluated in L929 cell lines, revealing a 95% acceleration in the wound healing process after 24 hours of exposure. Furthermore, the authors studied the photocatalytic activity of the ZnO-NPs by testing their ability to degrade methylene blue dye under solar irradiation, achieving 98% degradation after 240 min of sunlight exposure. In another study, Abdelaziz et al. [72] synthesized ZnO-NPs using a strain of *Penicillium expansum* and applied them to control *Fusarium oxysporum*, a phytopathogenic fungus that is the causal agent of wilt disease in *Solanum melongena L.* They observed antifungal activity *in vitro* throughout an agar well diffusion assay. Also, *in vivo*, the myco-synthesized ZnO-NPs suppressed *Fusarium* wilt disease in cultivated *Solanum melongena* L., reducing disease severity by 75% under pot conditions. Additionally, ZnO-NPs stimulated plant height, root length, fresh plant biomass, chlorophyll a and b content, total soluble carbohydrates, and other agronomic parameters. These findings suggest that myco-synthesized ZnO-NPs could serve as an effective alternative to synthetic fungicides, not only for eradicating *Fusarium* wilt in cultivated eggplant but also for enhancing plant growth. Finally, several reports have indicated that the biogenic synthesis of ZnO-NPs using endophytic fungi results in superior antimicrobial, antioxidant, and anticancer activities, as well as enhanced photocatalytic dye degradation efficiency compared to other synthesis methods [71].

Physical and Nutritional Factors Influencing the Myco-Synthesis of ZnO-NPs

The synthesis of NPs mediated by fungi may appear simple; however, numerous factors must be considered to control the size, shape, and stability of the ZnO-NPs. These factors include the culture conditions used to obtain the biomass or supernatant, the composition of the growth medium, temperature, pH, the amount of biomass or supernatant obtained, the type and concentration of the precursor, and importantly, the fungal strain used [72]. The carbon source in the growth medium can also influence NP synthesis. Although various reports indicate that enzymes are primarily responsible for mediating myco-synthesis, other biomolecules may also contribute to both the synthesis and capping of metal nanoparticles. To confirm this, one approach could involve performing the synthesis using purified enzymes and comparing the results with those obtained using crude enzyme extract. Another alternative could be the deletion of genes coding for these enzymes and determine if, even in their absence, the synthesis of the metal nanoparticles is carried out [72-73].

The pH of the synthesis medium affects the net charge of proteins and directly influences their solubility, aggregation and crystallization behavior [74]. It may also impact nanoparticle yield by altering the availability of functional groups involved in the reduction process [75]. Moreover, pH variations can modify the redox potential of functional groups present in the fungal filtrate. Under alkaline conditions, the high availability of OH^- ions facilitate a rapid nucleation process, leading to the formation of numerous small-sized nanoparticles [76]. Therefore, pH may affect not only nanoparticle synthesis but also capping efficiency and potentially their biological properties. Carboxylic acids, amides, esters, ethers and phenolic compounds mediate the synthesis of ZnO NPs [68].

Conclusion and Perspectives

Aflatoxin contamination in maize and other cereals in Argentina is a significant issue that affects both human health and food safety. Biological control strategies, such as the application of myco-synthesized ZnO-NPs, are essential to mitigate the associated risks and ensure safe food for the population. Among the available green synthesis methods for metal nanoparticles, myco-synthesis stands out as an eco-friendly approach for producing ZnO-NPs. Fungi offer adventages such as scalability, flexible cultivation, easy biomass processing and maintenance, efficient downstream handling, high yields of metabolites, extracellular redox proteins and enzyme production, tolerance to higher metal concentrations and economic viability. Therefore, fungal-synthesized nanoparticles can serve as eco-friendly fungicides, effectively inhibiting fungal growth and providing a sustainable alternative to chemical treatments.

Although numerous studies have successfully achieved the myco-synthesis of ZnO-NPs using various fungal species and have demonstrated their antifungal properties, no research has yet been investigated their potential to reduce mycotoxin production in crops. While the benefits of fungal-derived nanoparticles are considerable, challenges remain in scaling up production and ensuring consistent quality. Further research is necessary to optimize these processes for broader industrial applications, as well as to assess their environmental impact by conducting toxicological studies to confirm their minimal or null toxicity.

References

[1] Sharma, D., Kanchi, S., Bisetty, K. 2019. Biogenic synthesis of nanoparticles: a review. *Arabian Journal of Chemistry*, 12: 3576-3600. https://doi.org/10.1016/j.arabjc.2015.11.002.

[2] Sun, Q., Li, J., Le, T. 2018. Zinc oxide nanoparticle as a novel class of antifungal agents: current advances and future perspectives. *Journal of Agriculture and Food Chemistry*, 66(43): 11209-11220. https://doi.org/10.1021/acs.jafc.8b03210.

[3] Yadav, S., Mehrotra, G. K., Dutta, P. K. 2021. Chitosan-based ZnO nanoparticles loaded gallic-acid films for active food packaging. *Food Chemistry*, 334: 127605. https://doi.org/10.1016/j.foodchem.2020.127605.

[4] Seray M., Skender A., Hadj-Hamou A. S. (2020). Kinetics and mechanisms of $Zn^{2?}$ release from antimicrobial food packaging based on poly (butylene adipate-co-terephthalate) and zinc oxide nanoparticles. *Polymer Bulletin*, 78: 1021-1040. https://doi.org/10.1007/s00289-020-03145-z.

[5] Prasad, R., Bhattacharyya, A., Nguyen, Q. D. 2017. Nanotechnology in sustainable agriculture: recent developments, challenges, and perspectives. *Frontiers in Microbiology*, 8: 1014. https://doi.org/10.3389/fmicb.2017.01014.

[6] Sturikova, H., Krystofova, O., Huska, D., Adam, V. 2018. Zinc, zinc nanoparticles and plants, *Journal of Hazardous Materials*, 349:101-110, https://doi.org/10.1016/j.jhazmat.2018.01.040.

[7] Yadav, N., Kumar Garg, V., Kumar Chhillar, A., Singh Rana, J. 2023. Recent advances in nanotechnology for the improvement of conventional agricultural

systems: A review. *Plant Nano Biology*, 4:100032, https://doi.org/10.1016/j.plana.2023.100032.

[8] Akpomie, K. G., Ghosh, S., Gryzenhout, M. et al. 2021. One-pot synthesis of zinc oxide nanoparticles via chemical precipitation for bromophenol blue adsorption and the antifungal activity against filamentous fungi. *Scientific Reports*, 11: 8305. https://doi.org/10.1038/s41598-021-87819-2.

[9] Kairyte, K., Kadys, A., Luksiene, Z., 2013. Antibacterial and antifungal activity of photoactivated ZnO nanoparticles in suspension. *Journal of Photochemical Photobiology B: Biol*ogy 128, 78-84. https://doi.org/10.1016/j.jphotobiol.2013.07.017.

[10] Kumar, K., Shyamlal Kumar, S. B. R., Gupta, A., Mathur, M., Swami, A. K., Chaudhary, S. 2018. Efficacious fungicidal potential of a composite derived from nano-aggregates of Cu-Diclofenac complexes and ZnO nanoparticles. *Composites Communications*, 10: 81-88. https://doi.org/10.1016/j.coco.2018.07.004.

[11] Arciniegas-Grijalba, P. A., Patiño-Portela, M. C., Mosquera-Sánchez, L. P., Guerra Sierra, B. E., Muñoz-Florez, J. E., et al. 2019. ZnO-based nanofungicides: synthesis, characterization, and their effect on the coffee fungi *Mycena citricolor* and *Colletotrichum* sp. *Materials Science and Engineering: C*, 98: 808-825. https://doi.org/10.1016/j.msec.2019.01.031.

[12] Pena, G. A., Cárdenas, M. A., Monge, M. D. P., Yerkovich, N., Planes, G. A., Chulze, S. N. 2022. Reduction of *Fusarium proliferatum* growth and fumonisin accumulation by ZnO nanoparticles both on a maize-based medium and irradiated maize grains. *International Journal of Food Microbiology*, 363: 109510. https://doi.org/10.1016/j.ijfoodmicro.2021.109510.

[13] Mosquera-Sánchez, L. P., Arciniegas-Grijalba, P. A., Patiño-Portela, M. C., Guerra-Sierra, B. E., Muñoz-Florez, J. E., Rodríguez-Páez, J. E. 2020. Antifungal effect of zinc oxide nanoparticles (ZnO-NPs) on *Colletotrichum* sp., causal agent of anthracnose in coffee crops, *Biocatalysis and Agricultural Biotechnology*, 25:101579, https://doi.org/10.1016/j.bcab.2020.101579.

[14] Zaman, W., Ayaz, A., Park, S. 2025. Nanomaterials in Agriculture: A Pathway to Enhanced Plant Growth and Abiotic Stress Resistance. *Plants*, 14(5): 716. https://doi.org/10.3390/plants14050716.

[15] Mishra, P. K., Mishra, H., Ekielski, A., Talegaonkar, S., Vaidya, B. 2017. Zinc oxide nanoparticles: a promising nanomaterial for biomedical applications. *Drug Discov Today,* 22(12):1825-1834. https://doi:10.1016/j.drudis.2017.08.006.

[16] Hussain, I., Singh, N. B., Singh, A. et al. 2016. Green synthesis of nanoparticles and its potential application. Biotechnological Letters, 38: 545-560. https://doi.org/10.1007/s10529-015-2026-7.

[17] Sudheer S., Geetha Bai, R., Muthoosamy, K., Tuvikene, R., Kumar Gupta, V., Manickam, S. 2022. Biosustainable production of nanoparticles via mycogenesis for biotechnological applications: A critical review. *Environmental Research*, 204:111963, https://doi.org/10.1016/j.envres.2021.111963.

[18] United Nations (UN), 2024. Department of Economic and Social Affairs, Population Division (2024). World Population Prospects 2024: Summary of Results (UN

DESA/POP/2024/TR/NO. 9), undesa_pd_2024_WPP2024_Summary_of_Results. pdf. Accessed on May 2025.

[19] USDA, 2024. Corn | USDA Foreign Agricultural Service. United States Department of Agriculture, 1-13. Available from https://fas.usda.gov/.

[20] BCR, 2022. Bolsa de Comercio de Rosario. Available at: https://www.bcr.com.ar/es/mercados/gea/estimaciones-nacionales-de-produccion/estimaciones.

[21] Alaniz Zanon, M. S. A., Pena, G., Yerkovich, N. et al. Aflatoxins and fumonisins in maize under a climate change scenario. Biocontrol strategies at the pre-harvest stage. *European Journal of Plant Pathology*, 167: 551-567. https://doi.org/10.1007/s10658-023-02735-7.

[22] IARC (International Agency for Research on Cancer) Working Group on the Evaluation of Carcinogenic Risks to Humans 2002. Some traditional herbal medicines, some mycotoxins, naphthalene and styrene. IARC Monographs on the Identification of Carcinogenic Hazards to Humans.; 82:1-556. PMID: 12687954; PMCID: PMC4781602.

[23] Anwar, M. M., Aly, S. S. H., Nasr, E. H., El-Sayed, E. R. 2024. Correction: Improving carboxymethyl cellulose edible coating using ZnO nanoparticles from irradiated *Alternaria tenuissima*. *AMB Express*. 14(1):127. https://doi:10.1186/s13568-024-01779-0. Erratum for: AMB Express. 2022 Sep 7;12(1):116. doi: 10.1186/s13568-022-01459-x. PMID: 39543047; PMCID: PMC11564461.

[24] Hatab, M. H., Rashad, E., Saleh, H. M. et al. 2022. Effects of dietary supplementation of myco-fabricated zinc oxide nanoparticles on performance, histological changes, and tissues Zn concentration in broiler chicks. *Scientifical Reports*, 12(1):18791. https://doi:10.1038/s41598-022-22836-3.

[25] Shaat, S. K. K., Musleh, H., Zayed, H., Tamous, H., Issa, A., Shurrab, N. et al. 2019. Solution combustion-derived ZnO nanoparticles for photoanode of solar cells. *Materials Science and Engineering: B*, 241: 75-81, https://doi.org/10.1016/j.mseb.2019.02.008.

[26] Chandekar, K. V., Shkir, M., Badria M. Al-Shehri, S. et al. 2020. Visible light sensitive Cu doped ZnO: Facile synthesis, characterization and high photocatalytic response. *Materials Characterization*, 165: 110387, https://doi.org/10.1016/j.matchar.2020.110387.

[27] Little, D. J., Pfund, J. D., McLain, A. A., Sporie, J. A. et al. 2020. Synthesis of a Zinc Oxide/graphene hybrid material by the direct thermal decomposition of oxalate *Materials Research Express*, 7(6): 065005. https://doi:10.1088/2053-1591/ab9779.

[28] Jiang, Z., Laigui, B. L., Yuping Tong, Y., Yan, M. et al. 2023. Research progresses in preparation methods and applications of zinc oxide nanoparticles. *Journal of Alloys and Compounds*, 956: 170316, https://doi.org/10.1016/j.jallcom.2023.170316.

[29] Andreescu, S., Ornatska, M., Erlichman, J. S., Estevez, A., Leiter, J. C. 2012. Biomedical Applications of Metal Oxide Nanoparticles. In: Matijević, E. (eds) *Fine Particles in Medicine and Pharmacy*. Springer, Boston, MA. https://doi.org/10.1007/978-1-4614-0379-1_3.

[30] Yoon J and Oh, S. G. 2021. Synthesis of amine modified ZnO nanoparticles and their photocatalytic activities in micellar solutions under UV irradiation, *Journal of Industrial and Engineering Chemistry*, 96: 390-396, https://doi.org/10.1016/j.jiec.2021.01.043.

[31] Kumar, P., Kumar, A., Rizvi, M. A., Moosvi, S. K., et al. 2020. Surface, optical and photocatalytic properties of Rb doped ZnO nanoparticles. *Applied Surface Science*, 514: 145930, https://doi.org/10.1016/j.apsusc.2020.145930.

[32] Polte, J. 2015. Fundamental growth principles of colloidal metal nanoparticles–a new perspective. *Cryst Eng Comm*, 17(36): 6809-6830. https://doi.org/10.1039/C5CE01014D.

[33] Nath, S., Jana, S., Pradhan, M. & Pal, T. 2010. Ligand-stabilized metal nanoparticles in organic solvent. *Journal of Colloid and Interface Science*, 341(2):333-352. https://doi.org/10.1016/j.jcis.2009.09.049.

[34] Studart, A. R., Amstad, E., Gauckler, L. J. 2007. Colloidal stabilization of nanoparticles in concentrated suspensions. *Langmuir*. 23(3):1081-90. https://doi:10.1021/la062042s.

[35] Park, J. Y., Aliaga, C., Renzas, J. R., Lee, H., Somorjai, G. A. (2009). The role of organic capping layers of platinum nanoparticles in catalytic activity of CO oxidation. Catalysis letters, 129(1-2), 1-6. https://doi.org/10.1007/s10562-009-9871-8.

[36] Kvítek, L., Panáček, A., Soukupova, J., Kolář, M., Večeřová, R. et al. 2008. Effect of surfactants and polymers on stability and antibacterial activity of silver nanoparticles (NPs). *The Journal of Physical Chemistry C*, 112(15): 5825-5834. https://doi.org/10.1021/jp711616v.

[37] Niska, K., Knap, N., Kędzia, A., Jaskiewicz, M., et al. 2016. Capping agent-dependent toxicity and antimicrobial activity of silver nanoparticles: an *in vitro* study. Concerns about potential application in dental practice. *International journal of medical sciences*, 13(10):772-782 https://www.medsci.org/v13p0772.htm.

[38] Tejamaya, M., Römer, I., Merrifield, R. C., Lead, J. R. 2012. Stability of citrate, PVP, and PEG coated silver nanoparticles in ecotoxicology media. *Environmental science & technology*, 46(13): 7011-7017. https://doi.org/10.1021/es2038596.

[39] Asif, N., Amir, M. & Fatma, T. 2023. Recent advances in the synthesis, characterization and biomedical applications of zinc oxide nanoparticles. *Bioprocess and Biosystems Engineering*, 46: 1377-1398. https://doi.org/10.1007/s00449-023-02886-1.

[40] He, X., Deng, H., Hwang, H. 2019. The current application of nanotechnology in food and agriculture, *Journal of Food and Drug Analysis*, 27(1): 1-21, https://doi.org/10.1016/j.jfda.2018.12.002.

[41] Singh T. A., Das J., Sil PC. 2020. Zinc oxide nanoparticles: A comprehensive review on its synthesis, anticancer and drug delivery applications as well as health risks. *Advances in Colloid and Interface Science*, 286:102317. https://doi.org/10.1016/j.cis.2020.102317.

[42] Stracquadanio, C., Luz, C., La Spada, F., Meca, G., & Cacciola, S. O. 2021. Inhibition of mycotoxigenic fungi in different vegetable matrices by extracts of

Trichoderma species. *Journal of Fungi*, 7(6): 445. https://doi.org/10.3390/jof7060445.

[43] European Commission (EC) 2010. Commission regulation (EU) N° 165/2010 of 26 February 2010 amending regulation (EC) N° 1881/2006 setting maximum levels for certain contaminants in foodstuffs as regards aflatoxins. Official Journal of the European Union, L50, 8-12. https://www.fao.org/faolex/results/details/en/c/LEX-FAOC092874/.

[44] Agência Nacional de Vigilância Sanitária (ANVISA) (2011). Brazilian Sanitary Surveillance Agency: Resolução n°7, de 18 de fevereiro de 2011. https://anvisalegis.datalegis.net/action/UrlPublicasAction.php?acao=abrirAtoPublico&num_ato=00000007&sgl_tipo=RDC&sgl_orgao=RDC/DC/ANVISA/MS&vlr_ano=2011&seq_ato=000&cod_modulo=310&cod_menu=9882.

[45] Código Alimentario Argentino (CAA) 2021. Capítulo III: De los productos alimenticios. https://www. argentina.gob.ar/sites/default/files/anmat_capitulo_iii_prod_alimenticios.pdf.

[46] Sharma, D., Sharma, S., Kaith, B. S., Rajput, J., Kaur, M., 2011. Synthesis of ZnO nanoparticles using surfactant free in-air and microwave method. Applied Surface Science. 257, 9661-9672. https://doi.org/10.1016/j.apsusc.2011.06.094.

[47] Pena G. A., Cárdenas M. A., Monge M. P., Yerkovich N., Planes G. A., Chulze S. N. 2025. Effect of ZnO-nanoparticles on *Aspergillus flavus* growth and aflatoxin accumulation both on a maize based medium and irradiated maize grains." *World Mycotoxin Journal. Accepted.* Manuscript ID WMJ-1042.

[48] Camardo Leggieri, M. C., Toscano, P., & Battilani, P. 2021. Predicted Aflatoxin B_1 increase in Europe due to climate change: Actions and reactions at global level. *Toxins, 13*(4), 292. https://doi.org/10.3390/toxins13040292.

[49] Dinolfo, M. I., Martínez, M., Castanares, E., Arata, A. F. 2022. *Fusarium* in maize during harvest and storage: A review of species involved, mycotoxins, and management strategies to reduce contamination. *European Journal of Plant Pathology, 164*(2), 151-166. https://doi.org/10.1007/s10658-022-02548-0.

[50] Pickova, D., Ostry, V., Toman, J., & Malir, F. 2021. Aflatoxins: History, Significant Milestones, Recent Data on Their Toxicity and Ways to Mitigation. *Toxins*, 13(6): 399. https://doi.org/10.3390/toxins13060399.

[51] Mercosur 2002. MERCOSUR/GMC/Res. N◦ 25/02. Reglamento técnico MERCOSUR sobre límites máximos de aflatoxinas admisibles en leche, maní y maíz. Derogación de la Res. GMC N◦ 56/94. https://normas.mercosur.int/public/normativas/1047.

[52] Battilani, P., Barbano, C., & Piva, G. 2008. Aflatoxin B_1 contamination in maize related to the aridity index in North Italy. *World Mycotoxin Journal*, 1(4): 449-456. https://doi.org/10.3920/WMJ2008.x04.

[53] Benkerroum, N. 2020. Aflatoxins: Producing-molds, structure, health issues and incidence in Southeast Asian and Sub-Saharan African countries. *International Journal of Environmental Research and Public Health*, 17(4): 1215. https://doi.org/10.3390/ijerph17041215.

[54] Dövényi-Nagy, T., Rácz, C., Molnár, K., Bakó, K., Szláma, Z. et al. 2020. Pre-harvest modelling and mitigation of aflatoxins in maize in a changing climatic

environment-A review. *Toxins*, 12(12): 768. https://doi.org/10.3390/toxins 12120768.

[55] Ouadhene, M. A., Ortega-beltran, A., Sanna, M., Cotty, P. J. & Battilani, P. 2023. Multiple year influences of the aflatoxin biocontrol product AF-X1 on the *A. flavus* communities associated with maize production in Italy. *Toxins*, 1:1-15. https://doi.org/10.3390/toxins15030184.

[56] Broggi, L. E., Pacin, A. M., Gasparovic, A., Sacchi, C., Rothermel, A., Gallay, A., & Resnik, S. 2007. Natural occurrence of aflatoxins, deoxynivalenol, fumonisins and zearalenone in maize from Entre Rios Province. Argentina. *Mycotoxin Research*, 23(2): 59-64. https://doi: 10.1007/BF02946026.

[57] Garrido, C. E. Hernández Pezzani, C. Pacin, A. 2012. Mycotoxins occurrence in Argentina's maize (*Zea mays* L.), from 1999 to 2010, *Food Control*, 25(2): 660-665, https://doi.org/10.1016/j.foodcont.2011.11.043.

[58] Yu, J., Hennessy, D. A., Tack, J., & Wu, F. 2022. Climate change will increase aflatoxin presence in US Corn. *Environmental Research Letters*, 17(5): 054017. https://doi.org/10.1088/1748-9326/ac6435.

[59] Alaniz Zanon, M. S., Bossa, M., Chiotta, M. L., Oddino, C., Giovanini, D., et al. 2022. Pre-harvest strategy for reducing aflatoxin accumulation during storage of maize in Argentina. *International Journal of Food Microbiology*, 380: 109887. https://doi.org/10.1016/j.ijfoodmicro.2022.109887.

[60] Abdelkader, D. H., Negm, W. A., Elekhnawy, E., et al. 2022. Zinc oxide nanoparticles as potential delivery carrier: green synthesis by *Aspergillus niger* endophytic fungus, characterization, and *in vitro/in vivo* antibacterial activity. *Pharmaceuticals,* 15(9):1057. https://doi.org/10.3390/ph15091057.

[61] Rauwel, P., Küünal, S., Ferdov, S., Rauwel, E. 2015. A review on the green synthesis of silver nanoparticles and their morphologies studied via TEM. *Advances in Materials Science and Engineering*, 2015. Article ID 682749, 9 pages. http://dx.doi.org/10.1155/2015/682749.

[62] Prathna T. C., Mathew, L., Chandrasekaran, N., Raichur, A. M., Mukherjee, A. 2010. Biomimetic synthesis of nanoparticles: science, technology & applicability. In Biomimetics learning from nature. *Intechopen*. http://doi:10.5772/8776.

[63] Mohd Yusof, H., Mohamad, R., Zaidan, U. H., Abdul Rahman, N. A. 2019. Microbial synthesis of zinc oxide nanoparticles and their potential application as an antimicrobial agent and a feed supplement in animal industry: a review. *Journal of Animal Science and Biotechnology*, 10: 57. https://doi.org/10.1186/s40104-019-0368-z.

[64] Hulkoti, N. I. & Taranath, T. C. 2014. Biosynthesis of nanoparticles using microbes-a review. *Colloids and Surfaces B: Biointerfaces*, 121: 474-483. https://doi.org/10.1016/j.colsurfb.2014.05.027.

[65] Rafique, M., Sadaf, I., Rafique, M. S. & Tahir, M. B. 2017. A review on green synthesis of silver nanoparticles and their applications. *Artificial cells, nanomedicine, and biotechnology*, 45(7): 1272-1291. https://doi.org/10.1080/21691401.2016.1241792.

[66] Ahmed, J., Farhana, Manzoor, G. A. et al. 2025. *In-vitro* antifungal potential of myco *versus* bacteria synthesized ZnO NPs against chickpea and apricot pathogen. *Scientifical Reports*, 15: 148. https://doi.org/10.1038/s41598-024-84438-5.

[67] Zaki, S. A., Ouf, S. A., Albarakaty, F. M., Habeb, M. M., Aly, A. A., Abd-Elsalam, K. A. 2021. *Trichoderma harzianum*-Mediated ZnO Nanoparticles: A Green Tool for Controlling Soil-Borne Pathogens in Cotton. *Journal of Fungi*, 7: 952. https://doi.org/10.3390/jof7110952.

[68] Ghareeb, Y. R., Belal, E. B., El-Khateeb, N. M. M. et al. 2024. Utilizing biosynthesis of nanomaterials as biological agents for controlling soil-borne diseases in pepper plants: root-knot nematodes and root rot fungus. *BMC Plant Biology*, 24(1): 110. https://doi.org/10.1186/s12870-024-04760-y.

[69] Mekky, A. E., Farrag, A. A., Hmed, A. A., Sofy, A. R. 2021. Preparation of Zinc Oxide Nanoparticles using *Aspergillus niger* as Antimicrobial and Anticancer Agents. *Journal of Pure Applied Microbiology*, 15(3): 1547-1566. https://doi.org/10.22207/JPAM.15.3.49.

[70] Tamim, H. H., Zaghloul, R. A., Elameen, T. M., & Khalaphallah, R. 2024. Fungal synthesis of zinc oxide nanoparticles using *Aspergillus niger* for sustainable nanomaterial production and biological activity. *SVU-International Journal of Agricultural Sciences*, 6(4): 95-110. doi10.21608/svuijas.2024.317138.1389.

[71] Nehru, L., Kandasamy, G. D., Sekar, V., Alshehri, M. A., Panneerselvam, C. et al. 2023. Green synthesis of ZnO-NPs using endophytic fungal extract of *Xylaria arbuscula* from *Blumea axillaris* and its biological applications. *Artificial Cells, Nanomedicine, and Biotechnology*, 51(1): 318-333. https://doi.org/10.1080/21691401.2023.2232654.

[72] Abdelaziz, A. M., Salem S. S., Khalil A. M. A., et al. Potential of biosynthesized zinc oxide nanoparticles to control *Fusarium* wilt disease in eggplant (*Solanum melongena*) and promote plant growth. Biometals: An International Journal on the Role of Metal Ions in Biology, Biochemistry, and Medicine. 2022 Jun;35(3):601-616. doi: 10.1007/s10534-022-00391-8.

[73] Sumanth B., Lakshmeesha T. R., Ansari M. A., Alzohairy M. A., Udayashankar A. C., Shobha B., Niranjana S. R., Srinivas C., Almatroudi A. 2020. Mycogenic Synthesis of Extracellular Zinc Oxide Nanoparticles from *Xylaria acuta* and Its Nanoantibiotic Potential. *International Journal of Nanomedicine,* 15: 8519-8536. doi: 10.2147/IJN.S271743.

[74] Mousa, S. A., El-Sayed, E. S. R., Mohamed, S. S. et al. Novel mycosynthesis of Co3O4, CuO, Fe3O4, NiO, and ZnO nanoparticles by the endophytic *Aspergillus terreus* and evaluation of their antioxidant and antimicrobial activities. 2021. *Applied Microbiology and Biotechnology* 105, 741-753. https://doi.org/10.1007/s00253-020-11046-4.

[75] Rajan, A., Cherian, E., Baskar, G. 2016. Biosynthesis of zinc oxide nanoparticles using *Aspergillus fumigatus* JCF and its antibacterial activity. *International Journal of Modern Science and Technology*, 1: 52-57. ISSN: 2456-0235.

[76] El-Saadony, M. T., Alkhatib, F. M., Alzahrani, S. O., Shafi, M. E., Abdel-Hamid, S. E., Taha, T. F., Aboelenin, S. M., Soliman, M. M., Ahmed, N. H., 2021. Impact of mycogenic zinc nanoparticles on performance, behavior, immune response, and

microbial load in *Oreochromis niloticus*. *Saudi Journal of Biological Sciences*, 28(8): 4592-4604. https://doi.org/10.1016/j.sjbs.2021.04.066.

[77] Rafeeq, C. M., Paul, E., Vidya Saagar, E., Manzur Ali, P. P. 2021. Mycosynthesis of zinc oxide nanoparticles using *Pleurotus floridanus* and optimization of process parameters, *Ceramics International*, 47(9): 12375-12380, https://doi.org/10.1016/j.ceramint.2021.01.091.

[78] Rasha, E., Alkhulaifi, M. M., AlOthman, M., Khalid, I., Doaa, E., et al. 2021. Effects of Zinc Oxide Nanoparticles Synthesized Using *Aspergillus niger* on Carbapenem-Resistant *Klebsiella pneumonia In Vitro* and *In Vivo*. *Front Cell Infect Microbiol*. 16; 11:748739. doi:10.3389/fcimb.2021.748739. Erratum in: *Front Cell Infect Microbiol*. 2022 Feb 10; 12:843577. https://doi.org/10.3389/fcimb.2022.843577.

Index

Z